VIII International Scientific Practical Conference on Innovative Technologies in Engineering

IOP Conference Series: Materials Science and Engineering
Volume 221

Yurga, Russia
18 – 20 May 2017

ISBN: 978-1-5108-4511-4
ISSN: 1757-8981

Printed from e-media with permission by:

Curran Associates, Inc.
57 Morehouse Lane
Red Hook, NY 12571

Some format issues inherent in the e-media version may also appear in this print version.

Printed by Curran Associates, Inc. (2017)

For permission requests, please contact the Institute of Physics
at the address below.

Institute of Physics
Dirac House, Temple Back
Bristol BS1 6BE UK

Phone: 44 1 17 929 7481
Fax: 44 1 17 920 0979

techtracking@iop.org

Additional copies of this publication are available from:

Curran Associates, Inc.
57 Morehouse Lane
Red Hook, NY 12571 USA
Phone: 845-758-0400
Fax: 845-758-2633
Email: curran@proceedings.com
Web: www.proceedings.com

Table of contents

Volume 221

VIII International Scientific Practical Conference "Innovative Technologies in Engineering"

18–20 May 2017, Yurga, Russian Federation

Published online: 21 July 2017

Preface

Papers

IOP Conf. Series: Materials Science and Engineering **221** (2017) 011001 doi:10.1088/1757-899X/221/1/011001

PREFACE

Innovative Technologies in Engineering Conference Proceedings follow the same-name 8th International Scientific Practical Conference held at Yurga Institute of Technology (branch) of National Research Tomsk Polytechnic University from 18th to 20th May, 2017. The Proceedings contain the selected scientific reports submitted to the Conference. Having started in 1996 the scientific conference at Yurga Institute of Technology (branch) of National Research Tomsk Polytechnic University acquired its international status in 2010. This year scholars from Russia, Kazakhstan, Belarus, Ukraine, Azerbaijan and China have submitted their papers to the Conference.

The scientific reports published in these Proceedings have been revised and approved by the editorial team of the Conference. All of the reports exhibit clear, concise, and precise expositions that appeal to a broad international readership interested in mechanical engineering, welding, metallurgy, material science, as well as in computer-aided manufacturing. The reports present original ideas or results of general significance supported by clear reasoning and compelling evidence, and employ methods, theories and practices relevant to the research. The authors state clearly the questions and the significance of their research to theory and practice, describe how the research contributes to new knowledge; and provide tables and figures that meaningfully add to the narrative.

Organizing Committee of the Conference thank all the participants of the Conference for their fruitful work and personal contribution to the development of these Conference Proceedings.

Dmitry A. Chinakhov, PhD,
Conference Convenor,
Vice-Chancellor, Yurga Institute of Technology
of National Research Tomsk Polytechnic University

IOP Conf. Series: Materials Science and Engineering **221** (2017) 011002 doi:10.1088/1757-899X/221/1/011002

Peer review statement

All papers published in this volume of *IOP Conference Series: Materials Science and Engineering* have been peer reviewed through processes administered by the proceedings Editors. Reviews were conducted by expert referees to the professional and scientific standards expected of a proceedings journal published by IOP Publishing.

IOP Conf. Series: Materials Science and Engineering **221** (2017) 012001 doi:10.1088/1757-899X/221/1/012001

Investigation and Development of the Thermal Preparation System of the Trailbuilder Machinery Hydraulic Actuator

V Konev[a], E Polovnikov, O Krut, Sh Merdanov, G Zakirzakov

Tyumen Industrial University, Russian Federation
625000, Tyumen, St. Melnikaite 72, tel.: +7(3452)68-90-79

E-mail: [a]konevvv@tyuiu.ru

Abstract: It's determined that the main part of trailbuilders operated in the North is the technology equipped by the hydraulic actuator. Further development of the northern territories will demand using of various means and ways machinery thermal preparation, and also the machinery of the northern fulfillment. On this basis problems in equipment operation are defined. One of the main is efficiency supplying of a hydraulic actuator. On the basis of the operating conditions' analysis of trailbuilder hydraulic actuator operation it is determined, that under low negative temperatures the means of thermal preparation are necessary. The existing systems warm up only a hydraulic tank or warming up of the hydro equipment before the machinery operation is carried out under loading with intensive wears. Thus, with the purpose to raise the efficiency of thermal hydraulic actuator, operated far from stationary bases autonomous, energy saving, not expensive in creation and operation systems are necessary. In accordance with the analysis of means and ways of the thermal preparation of the hydraulic actuator and the thermal balance calculations of the (internal) combustion engine the system of the hydraulic actuator heating is offered and is being investigated. It contains a local hydraulic actuator warming up and the system of internal combustion engine heat utilization. Within research operation conditions of the local hydraulic actuator heating are viewed and determined, taking into account constructive changes to the local hydraulic actuator heating. Mathematical modelling of the heat technical process in the modernized hydraulic actuator is considered. As a result temperature changes of the heat-transfer and the hydraulic cylinder in time are determined. To check the theoretical researches and to define dependences on hydraulic actuator warming up, the experimental installation is made. It contains the measuring equipment, a small tank with the heat exchanger of the burnt gases of the internal combustion engine and the hydraulic cylinder with by-pass.

1. Introduction

Nowadays in the North of the Tyumen region and other northern regions of the country, a large fleet of trailbuilders is being operated. Recently, the equipment quantity used has sharply increased due to the completion of new gas fields in Yamal, large scale road construction and the laying out of new gas pipelines in Eastern Siberia. At the same time, the majority of this fleet consists of general purpose construction machinery, i.e. fleet not having the northern version [3]. It is obviously confirmed by the trailbuilder market expansion of rather cheap technique produced by South East Asia and China manufacturers. These manufacturers prefer to produce all-service equipment, shifting the problem of adaptation to local conditions onto the operating organizations.

The adaptation problems to northern conditions, thermal preparation of the internal combustion engine (ICE), hydraulic and hydromechanical transmissions of the systems of the hydraulic actuator of the trailbuilders are topical today. At the same time, if in the matters of the thermal preparation of ICE and hydraulic transmissions there has already been some progress, the thermal preparation of hydraulic systems has been less worked out [1].

2. Methodology

2.1. Analysis of the systems of thermal preparation of the hydraulic actuator

The most popular and having practical interest for today are the schemes of working fluid warming-up in the hydraulic systems of trailbuilders (Figure 1): throttle, reducing the hydro tank volume, electrical heaters, exhaust gases of internal combustion engines and with thermal accumulators usage.

Throttle warming of the working fluid (Figure 1a) allows in 16-25 minutes to come out to optimal thermal mode, at that practically does not require the introduction of construction changes into the hydraulic system. To install the throttle 3 is possible in the forcing line between a pump 2 and a distributor 4 or on the exhaust line behind the distributor in front of the filter. The latter is preferable, because in case of the throttle full closing the hydro system is defended from overloads by a safety valve of a distributor.

Very often in practice they use throttle warming-up by passing the fluid through the safety valve when stopping the operation equipment into insuperable obstacle, that allows to preheat the working fluid intensively. However, in this case the hydraulic actuator works under pressure, which is 15…20% higher than nominal, and under low temperatures because of inertance of transferring the direct impulse peak pressure is twice as high as nominal.

Such operation mode is accompanied by the vibration of a hydro system, increased noise, leads to a premature wear, and sometimes to a destroy of hydro equipment.

The noticeable affect is given by the usage of small hydro tanks (Figure 1b) for heating the working fluid. For instance, it reduces the duration of heating the hydraulic system of scrapers by 5 times, and other machinery by 1, 6…2 times. For the vehicles of northern version, especially having not heat intense hydraulic actuator, it's reasonable to use hydro tanks of decreased capacity, because only this small constructive change allows to raise the working capacity of a vehicle by 1,5…2 times in the heating period. When designing the hydro tank of a smaller capacity it must be remembered that the height of the fluid column in it should not be lower than the minimum level excluding the formation of a funnel in a sucking nozzle.

For the machinery with a heat intense hydraulic actuator the usage of small hydro tanks is also possible, however, in summer time it is necessary to use the tank of normal volume or a special oil cooler[2].

On (Figure1b) a simplified scheme to heat the working fluid of a hydraulic actuator is shown by changing the volume of a hydro tank, where: 1 – a hydro tank; 2 – a pump; 3 – a distributor; 4 – a small hydro tank, 5 – a switch.

Using the electric heating of the working fluid (Figure 1c) by heating elements, placed in the hydro tank is possible only if the self-moving machinery is equipped with the generator of the capacity not less than 30-40 % of the nominal pump arrangement capacity or feeding the machinery from the electrical-nets. In this case, as the estimates and the experience of foreign firms show it is possible to heat the working fluid from –40 °C to the optimal temperature for 15-20 minutes. Electrical heating of oil unlike other ways allows to perform local heating of hydro equipment, rather far away from a hydro tank and internal combustion engine. To exclude the local oil overheating and its chemical decomposition simultaneously with switching the electrical heaters on, a hydro pump must be switched on too, providing the continuous circulation of fluid in the hydraulic system. On (Figure 1c) the simplified scheme of the way to heat the working fluid with the help of electrical heaters is shown: where 1-a distributor; 2 – a pump; 3 – a hydro tank; 4 – an electric heating element.

IOP Conf. Series: Materials Science and Engineering 221 (2017) 012001 doi:10.1088/1757-899X/221/1/012001

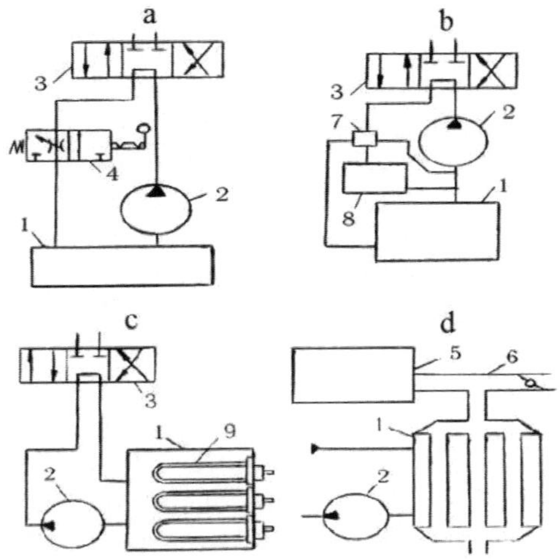

Figure 1: Schemes of working fluid warming-up: a) throttle, b) changing the hydro tank volume, c) electrical heating, d) using the heat of the exhaust gases

The biggest effect when heating the working fluid is reached by using the exhaust gases of ICE (Figure 1d), at that the duration of heating the working fluid is only 13-15 minutes, that is confirmed by experimental investigations. However, the complexity of the given way is in the fact that it is necessary to make significant changes in the construction of a hydro tank, making the technology of its production more complex and complicating the service in the process of operation [4]. But nevertheless this way found a wide implementation on earth-moving, load-lifting and other machines. On (Figure 1d) a simplified scheme to heat the working fluid with using the exhaust gases of ICE is shown: 1 –ICE; 2 – an exhaust tube; 3 – a hydro tank, 4 – a pump.

It should be noted, that the duration of hydraulic system transferring to the optimal thermal mode depends on the construction of hydraulic actuator as well as on the mode of machine operation, determined by its purpose. Thus, for example, at initial temperature of the working fluid of minus 40°C, it reaches the minimum value of optimal temperature of the working fluid of 0°C without using the devices for heating in a hydraulic system within 85 minutes, bulldozer-ripper after 125 minutes, and a scraper only after 5 hours of the excavator continuous work. That's why the choice of the way to heat the working fluid must take into account climatic operation conditions, and also type, design and technical level of the machine. For example, to raise the workability of hydraulic actuators during winter time in the southern regions and the regions of the middle part of the country, the usage of the tank of lower capacity may be enough, for the regions with cold climate it is also reasonable to use a smaller hydro tank but in connection with throttle or other kind of heating[7].

2.2. Description of the suggested system of thermal preparation of a machinery hydraulic actuator

The existing means of hydraulic actuator thermal preparation are power consuming, require much time, do not support its optimal thermal mode and local heating of the elements of the hydraulic actuator. Since the systems of thermal hydraulic actuator preparation that had been suggested before, implement the heating of oil, which is in the hydro tank and hydro line and moves in the small circle. At that hydro engines remain unheated and with multiple-viscosity oil in it significantly makes the beginning of movable elements activity harder in the places of compression.

It is necessary to reduce the load degree at the moment of first starting the hydro engine, which will allow reducing the wear intensity and prolonging the service duration of the hydraulic actuator elements [4]. It is also necessary to use the heat of internal combustion engine exhaust gases (heat utilization).

The necessity appears to optimize the hydro tank and layout of pipelines, and also heating the hydraulic actuator with local influence on its elements [5]. The existing schemes of heating the hydro systems do not provide autonomy of operation, have high energy intensity and labor intensity of service, limitation in implementing different arrangements of the hydraulic actuator – these lead to the decrease of ergonomic indexes of traibuilders.

Estimates of thermal balance of the internal combustion engine showed that nearly 25% of the thermal energy during engine operation comes out with exhaust gases (Figure 2).

In accordance with the analysis of thermal preparation means and the estimation of the thermal balance, the system of thermal preparation of the machinery hydraulic actuator is being developed. In the suggested design to reduce the time of thermal preparation a small hydro tank and local heating of the hydro cylinder are included additionally into the standard hydro system of the hydraulic actuator of trailbuilder machinery.

This system is based on the use of the thermal energy of the exhaust gases of the trailbuilder ICE. For this purpose, a gas switch is installed on the engine exhaust manifold which allows changing the exhaust gases stream into two directions: in the muffler (during the work) or via the gas pipeline, in the heat exchanger built in the hydro tank (while power fluid heating)[6, 8].

Connection of the gas pipeline with the heat exchanger is executed by means of the bellow valve which allows removal of the vibration loadings caused by the raised level of vibration of the trailbuilder ICE.

The heat exchanger is located in the lower part of the hydro tank so as to allow a reduction in the power fluid heating throughout the hydro tank, due to the increased convection of the power fluid; moreover, the heat exchanger placement below the level of the power fluid prevents it burning even in the case when the temperature of the wall of the heater is much higher than the boiling point of the power fluid.

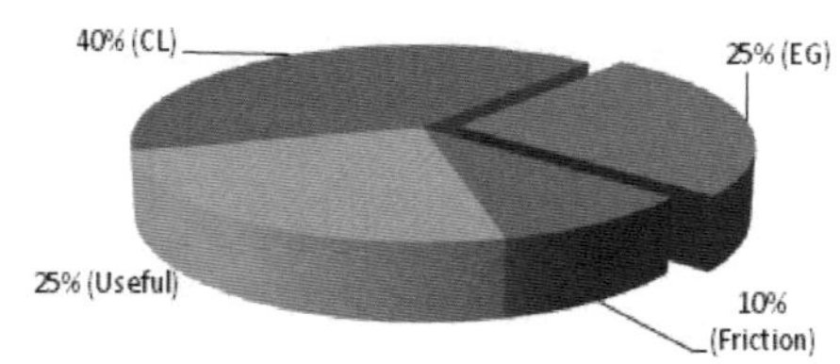

Figure 2: Thermal balance of the internal combustion engine. EG – exhaust gases CL – energy of cooling liquid. Friction, Useful energy

A key condition of modernization was the requirement for preservation of the commercial hydraulic system, including design of the hydro tank and the facilities for its attachment. It is necessary for the pump to be below the tank by not less than 50 cm and to provide delivery of warm power fluid to the pump from the hydro tank [10].

Calculations show that the time for a power fluid to heat from –40 °C to 0°C doesn't exceed 25 minutes; however, the heating time can be reduced if there is increased turbulence of the gas stream in the heat exchanger.

A small hydro tank is built between the main hydro tank and a hydro system pump. This allows before starting the hydraulic actuator to heat a small amount of the working fluid. At that the hydro tank is set lower than the main hydro tank and is connected with it by the tube of big section for the working fluid to come to a small hydro tank spontaneously. Also possible is the variant when a small tank is located in an optional place of the arrangement and is connected with the main tank by the tube of big section, additional pump for the viscous working fluid (tabular or spiral construction) and globe valve with valve control to doze the working fluid coming into a small hydro tank.

A pump serves to pump over the viscous working fluid into a small hydro tank. As opposed to the main pump (a hydro pump) it works at low pressure dozing the coming of the working fluid into a small hydro tank. In the construction the circulating pump is also provided for throwing a part of heated working fluid into the main tank to avoid the overheating of the working fluid. The block of automatic control serves to control the pneumatic valves and pumps, it collects the information from the temperature and fluid level sensors, located in the main hydro tank as well as in a small one for the automatics to throw out the excess of the heated fluid from the small hydro tank into the main one, and also to add new fluid into the small tank.

When the heating of the working fluid is not required, a valve, which is located in the exhaust system is closed. And with the help of three-way cock they transfer the hydro system onto normal mode of the main tank operation.

The local heating is authorized for the hydro equipment, located on the periphery of the machine. The hydro cylinder with the additional hydro line is being investigated [9]. By the branches stocked and rodless chambers of the hydro cylinder are connected with the additional hydro line through the valve or electromagnetic valve. The valve serves to combine the hydro cylinder chambers when heating and their division at normal hydro cylinder operation. The heating of the hydro cylinder is realized in the following way: movement of the heated in the hydro tank working fluid from one chamber into another by the additional hydro line, connecting stocked and rodless chambers of the hydro cylinder which provides the thermal change with a body, a stock, a piston of the hydro cylinder.

The open valve of the additional hydro line allows the heated working fluid to flow free in the chambers of the hydro cylinder. After finishing the cycle of hydro cylinder heating, the valve is closed thus stopping the movement of the working fluid through the bypass pipeline. When the valve is closed the working fluid acts on the piston, moving it in the body of the hydro cylinder. This gives the working mode to the hydro cylinder.

This allows heating the working fluid significantly quicker, than to heat the whole hydro tank of the machine, thus, reducing the time of heating preparation of the machine. The heated working fluid moves through the hydro pump, hydro equipment, pipelines, in this way the whole hydro system is heated, at that the system of heating hydro engines is provided by using the additional distribution of pipelines, at which the beginning of the working fluid movement is realized without stock movement (high wears in the beginning of operation are reduced).

2.3. Mathematical modelling process of the thermal preparation hydraulic actuator

To work out the given system it is necessary to carry out calculations, explorings, describing heat technological processes in a modernized hydraulic actuator.

For the by-pass hydraulic cylinder, according with one of the main laws of the fluid and gas mechanics – with the help of the equation of the thermal balance, of the motive medium (heat-transfer agent): time - total derivative from inner energy, given off mass of the motive medium is equaled to the sum of the heat power supplied or derived from heat-transfer agent and power of the inner viscous forces. One can usually disregard by heat flow in consequence of the heat conductivity of the heat-transfer agent along the normal in the entrance and outlet sections from the system of the heat preparation hydro cylinder. Heat power, transferred from heat-transfer agent (temperature T_a) to hydro cylinder (convective heat abstraction area $F_{a,}$ temperature T_a) is defined by the convective heat abstraction equation $Q_{ac} = \alpha_a F_a (T_a - T_c)$ [11].

Nonstationary processes in a heat-transfer agent system of the thermal preparation and in a hydro cylinder approximately are described by two differential equations for averaged temperatures of the heat-transfer agent T_a and hydro cylinder T_c, and also by the equation of the heat-transfer agent charges' balance in sections S_1, S', S", S_2:

$$c_a m_a \frac{\Delta T_a}{\Delta t} = G_1 c_a T_1 - c_2 c_a T_2 - G' c_a T' + G'' c_a T'' - Q_{ac} \quad (1)$$

$$c_c m_c \frac{\Delta T_c}{\Delta t} = Q_{ac} - Q_в \quad (2)$$

$$G_2 = G_1 - G' + G'' \quad (3)$$

where Q_{ac}= -Q_{ca}- heat power, supplied (Q_{ac}>0) from the heat-transfer agent to hydro cylinder by the heating of the hydro cylinder by the heat-transfer agent $Q_{в} = Q_{n}$.

Starting conditions are written down as: $t = t_0$, $T_a = T_{a0}$, $T_c = T_{c0}$.

Boundary conditions are set for the heat-transfer agent temperature in the section $S_{1:}T_1 = T_{10}(t)$. On every temporal step from $t_1 = t^{(n)}$ до $t_2 = t_1 + \Delta t = t^{(n+1)}$ are considered known such charges as G_1, G', G'', and also parameters c_a, m_a, c_c and m_c.

Besides heat ratings Q_{ac} and $Q_{в}$ can be determined by known parameters of the convective heat exchange from the heat-transfer agent to hydraulic cylinder $A_c = \alpha_a F_a$ and from the hydraulic cylinder in the environment $A_в = \alpha_в F_c$; $Q_{ac} = A_c(T_a - T_c)$; $Q_в = A_в(T_c - T_в)$.

The system of equations is solved by the standard methods, for example, by Runge-Kutta method. As a result it is found the temperature changing of the convective heat exchange and the hydraulic cylinder in time $T_c = T_c(t)$, $T_2 = T_2(t)$. This result is the aim of studying thermal preparation hydraulic cylinder to start.

To warm up by-pass hydraulic cylinder there are two ways of working fluid:

1. Through by-pass pipe into the rod end.
2. Setting up pressure in the cylinder cavity it is necessary to press the plunger, pulling out a rod.

In the first case, fluid needs to deal with resistance in the pipe and discharge resistance (from hydraulic cylinder rod end to tank). In the second case, discharge resistance reacts against fluid in the rodless cavity. Because of changing in the hydraulic machinery system it is made up a calculation of the hydraulic losses, and it is compared with working pumping pressure, applied in the hydraulic system. Pressure losses in the hydraulic system is determined taking into account machine working in the conditions of low temperature (t = –40 °C).

Pressure losses in hydrosystem are accounted for the fluid friction on piping walls and hydro equipment, and also inner fluid friction. Losses are up to the length, the diameter and forms of the pipe - lines, flow velocity and viscosity of working fluid, hydro system branching, fluid flow regime in the pipe-line.

Overall pressure losses in hydro system (ΔP) are defined as a sum of the pressure losses to overcome hydraulic resistance on the friction with the length (ΔP_L), local resistance (ΔP_M), and also pressure losses in the hydro equipment ΔP_H and is defined on formula:

$$\Delta P = \Delta P_L + \Delta P_M + \Delta P_H \tag{4}$$

Pressure in a system has to satisfy condition:

$$P_{nom} \leq P_{nom} + \Delta P \leq P_{max} \tag{5}$$

where P_{nom} – is a nominal pumping pressure, *MPa*; P_{max} – is a maximum pumping pressure, *MPa*

If the condition is met, so offered modernization doesn't influence on the normal work of the hydraulic system. To corroborate theoretical explorings it is necessary to carry out experimental explorings using empirical model and to compare derived results with the control data[10].

Theoretical explorings showed, that reactions in the hydraulic cylinders and pressure losses in a hydro system with temperature of the working fluid under heating (40 °C) and after heating (+50 °C) provide piston immobility under warming up of the by-pass hydraulic cylinder.

Timing warming up of the hydraulic cylinder with the help of the working fluid showed duration 5-10 minutes in the temperature range of the surrounding air from –40 °C till –15 °C. At the same time it is defined temperature lowering in a hydraulic tank on the warming-up stage of hydraulic cylinder.

It is necessary to define the heat quantity which is useful for a small hydraulic reservoir. Heat quantity is an energy, which body loses and gets by the heat transfer. Being getting cold the hydraulic cylinder will lose some amount of heat, under heating – absorb. Heat quantity is up on the next values. In the first place, the more the hydraulic reservoir tank, the more heat quantity it is necessary to use to change its temperature on one degree. In the second place, heat quantity which is useful for the hydraulic tank warming up is up on the heat capacity of the working fluid. Thirdly, temperatures' difference of the working fluid before and after heat-transmission influences on the total heat amount which is useful to warm up hydraulic reservoir:

$$Q = c \cdot m \cdot (t_2 - t_1) \qquad (6)$$

where Q – is a heat quantity (J), m – is a mass of the working fluid of the hydro reservoir (kg), c – specific heat of the working fluid (J/kg °C), (t_2-t_1) – is a difference between initial and final temperatures of the working fluid (°C).

This formula helps to make heat quantity calculations, which is useful to warm up the hydro tank or how much heat hydro tank gives off by cooling.

Figure 3 presents results of calculations for small hydro tank (volume 10 liters)

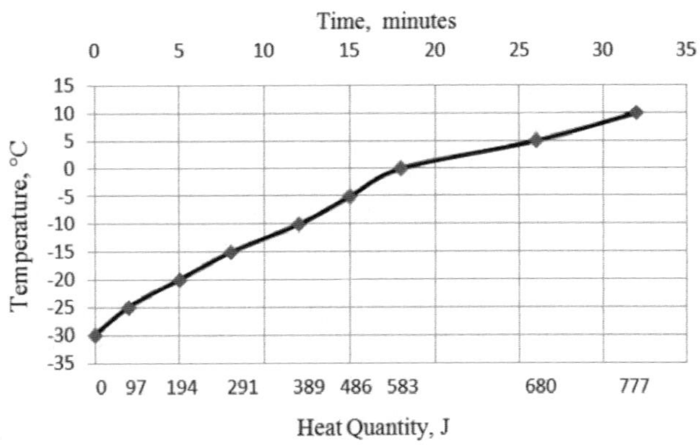

Figure 3: Heat quantity of the small hydro tank warming up

3. Experimental investigations

The aim of experimental investigations concludes theoretical researches and dependences' definition to warm up hydraulic drive.

Principle diagram which is worked out for hydro system of the experimental installation is shown in Figure 4.

Hydro cylinders (1 and 2) are managed by the separate distributors (7). Characteristics' measurements of working fluid are carried out inlet and outlet hydraulic cylinder, and also hydro tank.

It's important to fix such parameters as temperature and pressure in hydraulic cylinders. It allows getting the more clear performance about processes' characteristics. Thereupon there was set a distributing connecting pipe with measuring equipment in every hydraulic cylinder, in a pressure opening.

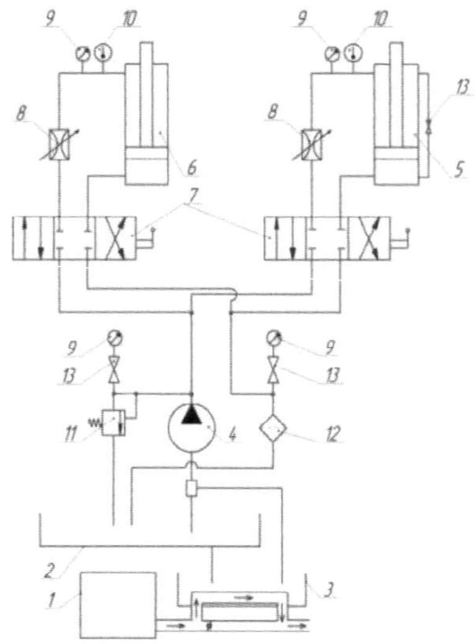

Figure 4: Hydraulic scheme of the experimental installation: 1 – internal-combustion engine, 2 – tank, 3 – a small hydro tank, 4 – pump, 5 – modernized hydraulic cylinder; 6 – control hydraulic cylinder; 7 – hydraulic control valve, 8 – throttle, 9 – manometer, 10 – temperature sensor, 11 – by-pass valve, 12 – filter, 13 – valve

Figure 5 shows a laboratory pattern of the by-pass hydraulic cylinder. To accelerate the working fluid warming up in the hydro tank, there was worked out a construction to use a small hydro tank (Figure 6). Such construction allows considerably to decrease working fluid time-warming up, it's very important for technics, which is working by low negative temperatures.

Figure 5: By-pass hydraulic cylinder

Figure 6: A small hydraulic tank with heat-exchanger

By warming up of hydraulic cylinder with by-pass pipeline, the analysis of warming up diagrams showed: the wall warming up of the experimental cylinder is carried out quicker having –20°C and –40°C accordingly (10.5 and 19 minutes), than in case of control hydraulic cylinder (12.5 minutes and 24 minutes), warming up character – at the beginning of the warming up we can see the acceleration, later stable, smooth. The warming up of the control hydraulic cylinder is carried out evenly [11].

To investigate hydro tank warming up, which uses the heat of the used gases of the internal-combustion engine, there was set a heat-exchanger in the lower part of a small hydro tank. Through

IOP Conf. Series: Materials Science and Engineering **221** (2017) 012001 doi:10.1088/1757-899X/221/1/012001

this heat-exchanger was run a hot stream of used gases, warming up working fluid of a small hydro tank (Figure 6).

In the connective pipe of the modernized hydraulic cylinder temperature of the working fluid is growing quickly, and tries to reach temperature of working liquid in the hydraulic tank. On this reason there was a circulation of working fluid on the additional hydro line. Temperature rises till 0°C for 0.75 minutes. In the connective pipe of the control hydraulic cylinder temperature of working fluid rises more slowly, than modernized hydraulic cylinder, but more stable. Temperature rises till 0°C for 3 minutes (Figure 7, Figure 8).

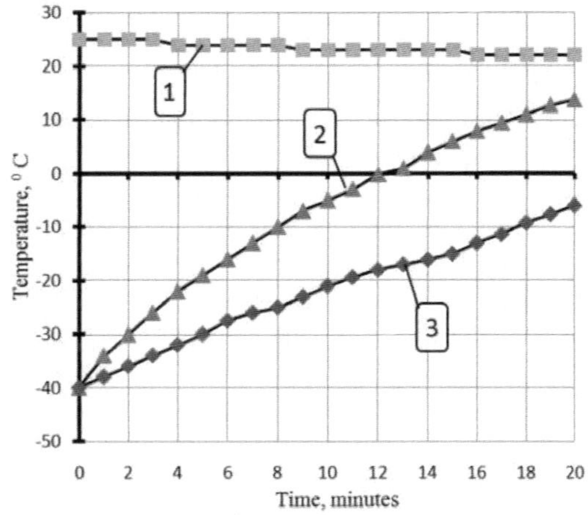

Figure 7: Diagram of control hydraulic cylinder warming up $t_{o.в.}$ = –40 °C:
1 – hydro tank, 2 – connective pipe, 3 – hydro cylinder

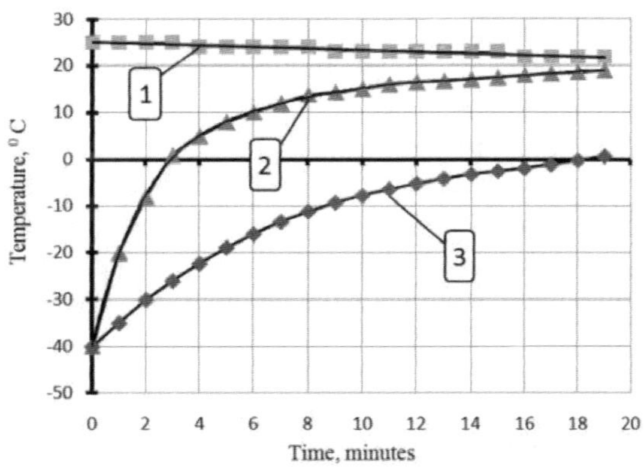

Figure 8: Diagram of the modernized hydro cylinder warming up $t_{o.в.}$ = –40 °C:
1 – hydro tank, 2 – connective pipe, 3 – hydro cylinder

During experiments readings were taken from measuring instruments, processing of measurements' results is given in the diagram of a small hydro tank (Figure 9) (by $t_{o.c.}$ = –28 °C).

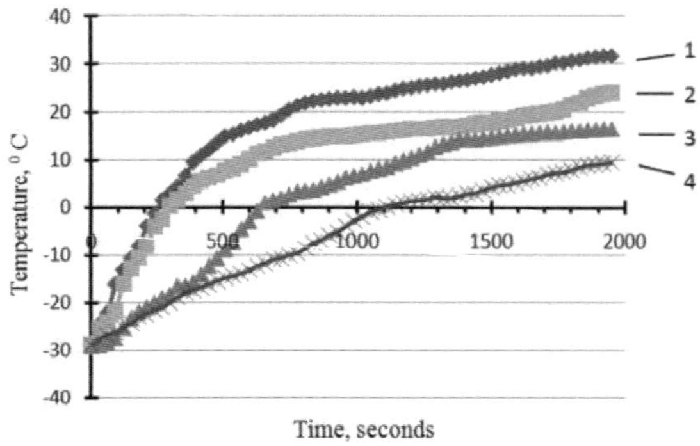

Figure 9: Diagram of the warming of the small hydro tank. Sensors are set in a hydro tank: 1 – inlet heat-exchanger, 2 – middle heat-exchanger, 3 – outlet heat-exchanger, 4 – middle-volume hydro tank

The diagram shows that the highest temperature is inlet heat-exchanger in the hydro tank. The lowest temperature of the working fluid - is in the middle-volume hydro tank. At the same way temperature growth dynamics of the working fluid in the hydro tank also differs. The most intensive warming up is inlet-hydro tank, and uniform warming up is in the middle-volume hydro tank.

Conclusion

The majority of the hydraulic systems' break-downs of trailbuilders, operated in the North are connected with an influence of low negative temperatures. Therefore there is a need to apply machines of the northern fulfillment, working fluids to use during the winter period and means of pre-starting thermal preparation of a hydraulic actuator.

The analysis of the earlier offered systems of a hydraulic actuator thermal preparation showed the following shortcomings: throttling – big resistance, low coefficient of efficiency, a warming up by burnt gases of the internal combustion engine and energy warming up – construction complexity and the construction changing of a regular hydro tank. All systems warm up only the main hydro tank. A large amount of energy and time during the start of a hydraulic actuator in work is spent for it. Thus, local warming up of the hydro equipment (hydraulic cylinders) elements at the beginning of work of the machine hydraulic actuator isn't carried out. It conducts to intensive wears, especially at the time of the beginning of work and to decreasing in term of equipment operation.

In the offered construction, to decrease the thermal preparation time and means for the modernization of the existing technology (without changing a regular hydro system), a small hydro tank and local hydraulic cylinder warming up are included extra in a regular hydro system of the trailbuilders. It allows heating the working fluid significantly quicker, than to heat the whole hydro tank of the machine. The heated working fluid moves through the hydro pump, hydro equipment, and pipelines. In this way the whole hydro system is heated. At the same time the system of heating hydro engine is provided by using the additional distribution of pipelines, at which the beginning of the working fluid movement is realized without stock movement.

The system of equations is solved by the standard methods, for example, by Runge-Kutta method. This result is the aim of studying thermal preparation hydraulic cylinder to start. Determination of the heat amount, which is necessary to warm up a hydro tank depends on the hydro tank mass, thermal capacity of the working fluid, temperatures' difference of the working fluid before and after a heat-transfer.

Means of the warming up of the working fluid should be chosen taking into account climatic conditions of operation, type and construction of the trailbuilder hydraulic actuator, technical production rate. For example, to increase the working capacity of the hydraulic actuator during the winter period in the southern areas and regions of a midland of the country there can be enough the application of a smaller capacity tank, with frigid climate it is also expedient to apply the smaller hydro tank, but in combination with the heat exchanger of the burnt gases or other means of a warming up.

References

[1] Anisimov, I., Ivanov, A., Chikishev, E., Chainikov, D., Reznik, L. Assessment of gas cylinder vehicles adaptability for operation at low ambient temperature conditions, WIT Transactions on Ecology and the Environment, 190 VOLUME 1, pp. 685-695, 2014.

[2] Chmil V.P. Hydropneumatic drive / Chmil V.P. - SPb .: SPbGASU, 2010. - 176 with.

[3] Chowdou, O.A. Development of a methodology for assessing the effect of climatic conditions on the operation of road-building machines: author's abstract. Diss. ... cand. Tech. Sciences / OA. Chowdou. - St. Petersburg, 2009. 128 p.

[4] H. Exner, R. Freitag, Dr. Qays H. Lang R., J. Oppolzer, P. Schwab, E. Sümpfen, W. Ostendorff, M. Raik // Lehrgang Hydraulik. Band 1. Rexroth Bosch-Gruppe: Bosch Rexroth AG Service Automation Didaktik der Erbach, Deutschland, 2003. 323 p.

[5] Karnaukhov N.N. Adaptation to the conditions of construction machinery in the Russian North and Siberia. – Moscow: Nedra, 1994 352.

[6] Kaverzin S.V. Performance hydraulic drive at low temperatures / Kaverzin S.V. – Atlanta: Publishing House of Krasnoyarsk University, 1986. 144.

[7] Konev V., Borodin D., Polovnikov E. Development of a universal data acquisition system with control functions based on analog to digital converter: Scientific Enquiry In The Contemporary World: Theoretical Basics And Innovative Approach 2016. 220 pp. 197-200 pp. ISBN-10:1-941655-35-1

[8] Konev V., Merdanov Sh., Karnaukhov M. & Borodin D. Thermal preparation of the trailbuilder fluid drive, WIT Transactions on Ecology and the Environment, 190 VOLUME 1, pp. 697-706, 2014.

[9] Merdanov Sh.M., Jakubowski Y.Y., Konev V.V., Karnaukhov M.M., Building and road machines. Research and development of training systems thermal hydraulic drive road construction machinery Moscow, № January, pp. 27-29, 2013.

[10] Vashurkin I.O. Thermal preparation of construction machinery in the harsh climate. - St. Petersburg.: Nauka, 2005. 238 p.

[11] Vasilchenko V.A. Features of operation of mining machines with a hydraulic drive at low temperatures / V.A. Vasilchenko / Mining Industry, 2 2006 URL: www.mining-media.ru/en/article/newtech/1151-osobennosti.

IOP Conf. Series: Materials Science and Engineering **221** (2017) 012002 doi:10.1088/1757-899X/221/1/012002

Dynamic Action of the Shielding Gas Jet Upon the Process of Consumable Electrode Welding

D A Chinakhov
Yurga Institute of Technology of Tomsk Polytechnic University
652055, Russia, Kemerovo region, Yurga, ul. Leningradskaya 26

E-mail: chinakhov@tpu.ru

Abstract. The author considers the gas-dynamic action of the shielding gas jet during consumable electrode welding on the processes in the welding area and properties of the weld joints produced from alloyed high-strength steel 30HGSA. The results of the research on the gas-dynamic action of the jet on the process of consumable electrode welding are provided. It was established that the method of consumable electrode welding with double-jet gas shielding ensures regulated gas dynamics in the welding area and allows controlling the electrode metal transfer, the chemical composition of the weld metal, stabilizing the process of welding and ensures stable high mechanic properties of the weld joints.

Introduction.

At the present time welding is applied for producing permanent joints of the widest range of metallic, nonmetallic and composite constructional materials under the conditions of earth atmosphere, World's ocean and aerospace. In spite of continuously growing application of light metal alloys, polymer and composite materials in construction, steel remains the basic constructional material. Consumable electrode welding in the atmosphere of shielding gases stays one of the main production methods in the process of developing the material basis of modern civilization.

The history of the thermal processes determines the productivity of the basic metal and welding filler fusion, trend and completeness of the metallurgic processes in the weld pool, conditions of weld and heat-affected zone metal structure formation, performance characteristics of the weld joints [1-5].

The shielding gas is normally used to protect the welding area from the harmful action of the air but, under certain conditions, it can become the instrument for controlling the processes in the weld zone and the properties of weld joints. A number of studies [6-13 and other] investigating the influence of the shielding gas flow upon the process of consumable electrode welding notice that increase in the gas flow rate is associated with improvement of the quality of the weld zone protection, weld formation, welding process stabilization, improvement of the weld metal properties. Increase of the shielding gas flow ensures increased hardness of the gas jet which is especially important when completing the welding works in an exposed position [2, 6, 10].

Methodology.

Significant change of the dynamics of shielding gas (Fig. 1) results in changing processes taking place in the weld zone (transfer of electrode metal droplets, metallurgic processes in the drop, thermal processes, dynamics of the weld pool, etc.) and, as a result, changing the performance characteristics of the weld joints [7, 14].

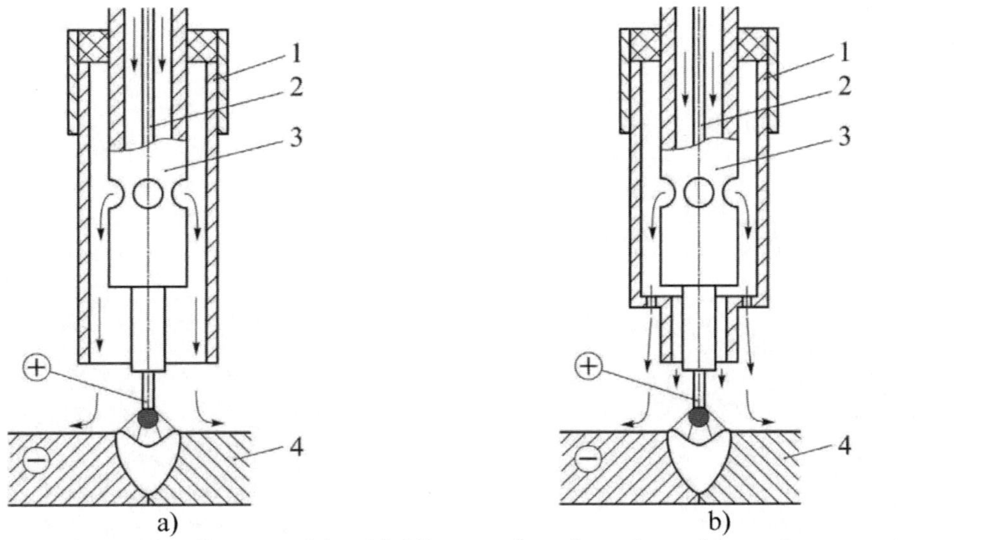

Fig. 1. The diagram of the shielding gas flow from the weld nozzle:
a) traditional single-jet gas shielding; b) double-jet gas shielding: 1 – the nozzle; 2 – the electrode wire; 3 – the contact tip;4 – the part

Action of the shielding gas jet upon the electrode metal droplet is 12 times more efficient under the double-jet gas shielding than under the single-jet one (Fig. 2).

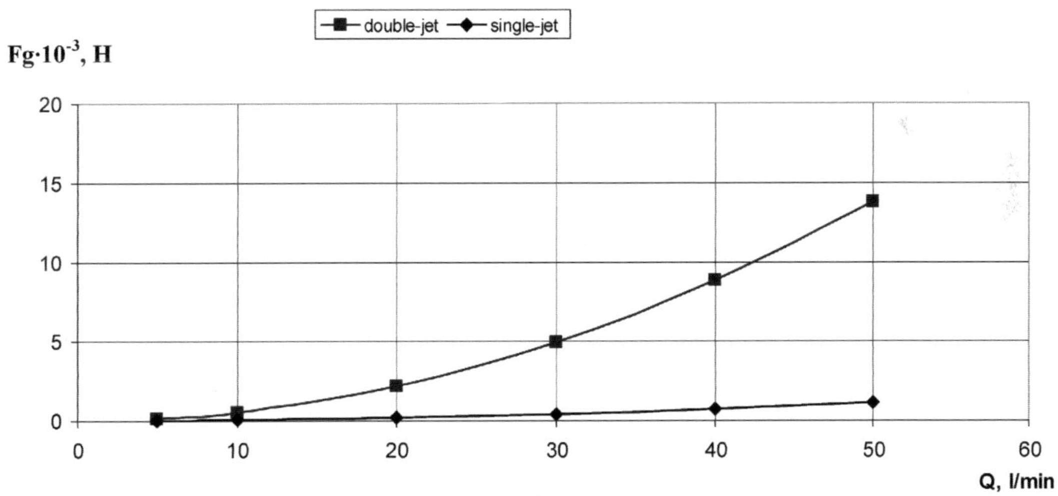

Fig. 2. The diagram of the changes of the efficiency of shielding gas jet action depending upon the gas flow rate

With the increase of the shielding gas flow acting upon the electrode metal droplet under the double-jet gas shielding we observe growth of frequency and stability of droplets transfer into the weld pool as the droplet takes up the coaxial position with the electrode and the chaotic oscillations of the droplet reduce.

The conditions of the experiment: automated single-pass welding of the plates from 30HGSA steel completed with Sv-08G2S welding wire 1.2 mm in diameter in CO_2 with stationary arc downhand with single-jet (traditional) and double-jet (developed), welding current I= 195…200 A, electrode wire stick-out distance L = 12 mm, shielding gas flow rate = 20 l/min, arc voltage U = 26…27 V, welding speed V = 25 cm/min.

Results and discussion.

The results of analysis of oscillograms and frames of high-speed video recording of the experimental studies (Fig. 3) showed that frequency of droplet transfer increases upon the average by 1.6 under double-jet shielding as compared to the single-jet one, together with reduction of the droplet size from the mean of 2.4 mm to 2.00 mm.

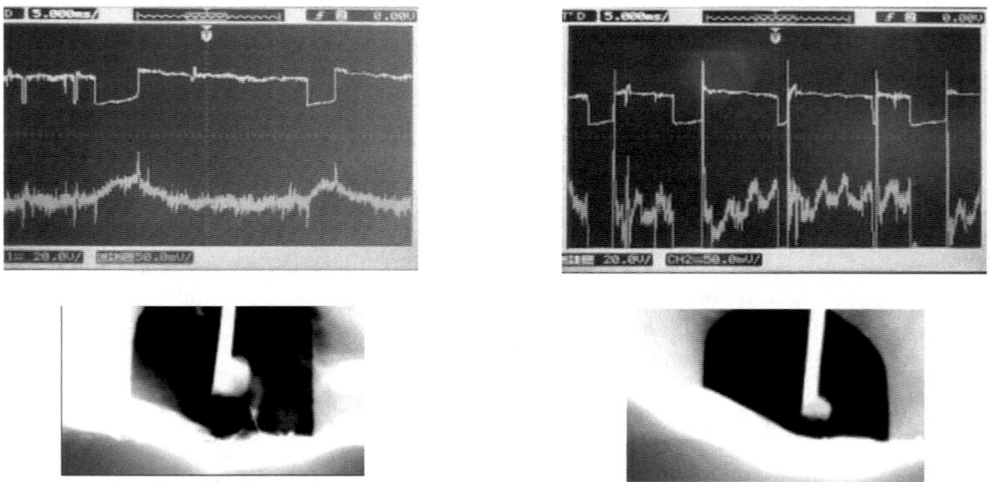

a) b)

Fig. 3. Oscillograms and frames of high-speed video recording:
a) traditional single-jet gas shielding; b) double-jet gas shielding

As the gas-dynamic action of active shielding gas jet upon the processes in the weld area increases the controlled change of the chemical composition and structure of the weld metal [7, 15] and, as a result, change of the performance characteristics of the weld joints also take place. The dynamics of the weld pool changes as well and here two aspects should be considered and taken into account (Fig. 4). First, the direct action of the active shielding gas upon the liquid metal of the weld pool around the welding arc. It results in growth of the liquid layer under the arc as there is no way out for the molten metal in the volume limited by solid metal and it starts flowing under the arc and rising it (Fig. 5 b). Growth of the liquid layer (b) results in reduction of the penetration depth (h).

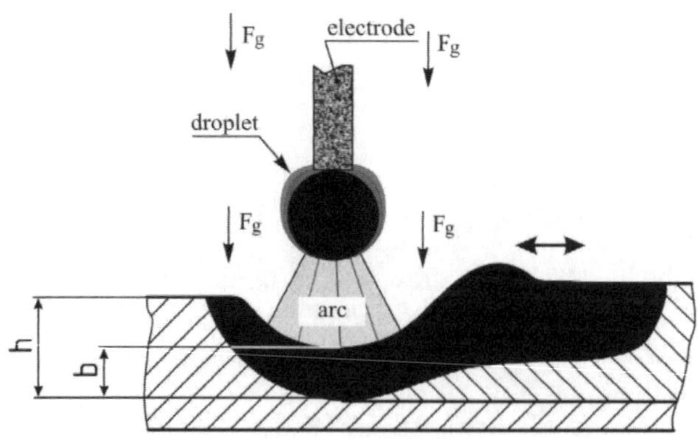

Fig. 4. The diagram of the jet action upon the weld pool under the double-jet gas shielding

In the second case, with increase of the shielding gas action upon the electrode metal droplet under the double-jet gas shielding the stability and frequency of droplet transfer into the weld pool grow and the size of the droplets reduces. Increase of the frequency of droplet transfer leads to reduction of time between short circuits, i.e. the workpiece heating time reduces and so does the penetration depth. The reproducibility and predictability of the chemical composition and properties of the weld increase.

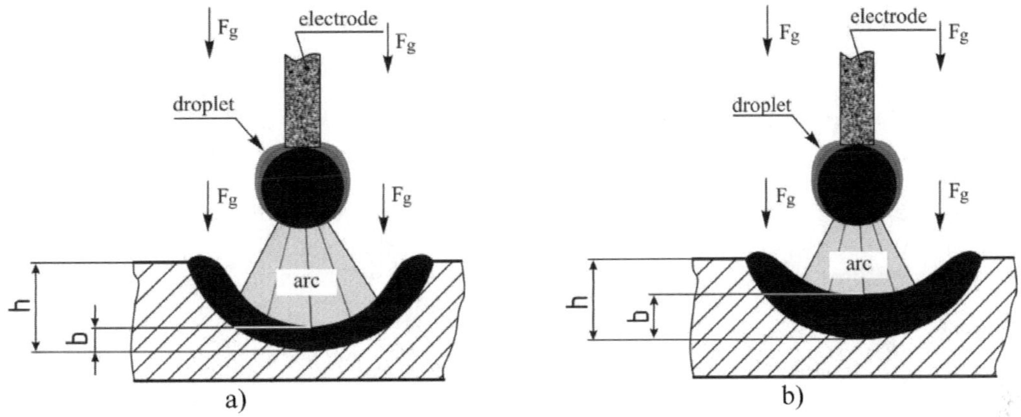

Fig. 5. The diagram of the changes of penetration depth and size of the liquid layer under the arc in the weld pool: a) single-jet gas shielding, b) double-jet gas shielding

In Table 1 we present the mechanic properties of 30HGSA steel samples welded with the welding methods under study. It is obvious that the welding method with application of double-jet gas shielding ensures more stable, better mechanic properties. The metal of the weld becomes stronger with some reduction of plasticity which is an indirect indicator of more intense stirring of metal in the weld pool [7, 15].

Table 1. The results of mechanic tests of the welded samples produced from 30HGSA steel

Welding method	Ultimate tensile strength, MPa	Flow limit, MPa	Impact hardness, J/cm2 under +20 °C, notched		Hardness	
			Along the weld center	Along HAZ	Weld, HRB	HAZ, HRC
Single-jet	540…640	435…535	138…150	121…163	86…90	22…23
Double-jet	750…790	590…610	88…126	106…176	93…98	24…26

Chemical analysis of the weld metal was completed on the welded samples (Table 2)

Table 2. Chemical analysis of the basic metal, welding wire and weld metal of the welded samples

Material	Mass fraction of the elements, %				
	C	Mn	Si	Cr	Ni
Basic metal 30HGSA	0.28…0.34	0.8…1.1	0.9…1.2	0.8…1.1	–
Welding wire Sv-08G2S	0.05…0.11	1.8…2.1	0.7…0.95	≤0.2	≤0.25
Single-jet	0.18	1.29	0.78	0.52	0.12
Double-jet	0.18	1.2	0.75	0.57	0.12

When welding with application of double-jet gas shielding the content of chromium increases by 0.05%. This indicates that when welding with double-jet gas shielding the stirring of the basic and electrode metal in the weld pool is more intense. The amount of manganese is reduced by 0.09% and that of silicon – by 0.03%. The explanation for this fact is that more intense gas-dynamic action on the droplet of molten electrode metal when welding with application of double-jet gas shielding results in increased frequency of droplet transfer, reduction of their size and growth of their surface activity [7, 15]. When changing the gas-dynamic action we can control the electrode metal droplets transfer, the chemical composition of the weld metal, thermal and other processes of consumable electrode welding and, this way, form the required properties of the weld joints.

Conclusion.

We have established that when completing consumable electrode welding in CO_2 the rate of the gas flowing from the welding nozzle has a significant impact upon the processes in the weld zone and upon the characteristics of the weld joint. The method of consumable electrode welding with application of double-jet gas shielding ensures regulated gas dynamics in the weld area and allows controlling electrode metal transfer, chemical composition of the weld joint metal, stabilizing the welding process, ensures stable high mechanic properties of the weld joints.

The research was completed with the financial support of Russian Science Foundation (project 16-19-10010).

References.

[1] Welding and welded materials: in 3 vol. Vol. 1. Weldability of materials: reference edition / ed. by E.L. Makarov. Metallurgiya, 1991. 528 p.

[2] Consumable electrode welding of steels in the shielding gases. Engineering and technology of the future// A.G. Potapievsky, Yu.N. Saraev, D.A. Chinakhov; Yurga Institute of Technology. Tomsk Polytechnic University publishers, 2012. 208 p.

[3] Shorshorov M.H. General metallurgy of steel and titanium alloys welding. Nauka. 1965. 337p.

[4] Livshits L.S., Hakimov A.N. Metal science of welding and thermal treatment of weld joints. 2nd edition, revised and completed. Mashinostroyeniye. 1989. 339 p.

[5] Lahtin Yu.M. General metallurgy and thermal treatment of metals. 3d edition, revised and completed. Metallurgiya. 1983. 359 p.

[6] Novozhilov M. N. The basic principles of the metallurgy of arc welding in the gases. Mashinostroyeniye. 1979. 231 p.

[7] Chinakhov D.A., Zuev A.V., Filimonenko A.G. Gas-dynamic Impact of a Shielding Gas Jet on the Drop Transfer When Welding with a Consumable Electrode. Advanced Materials Research. 1040 (2014) pp. 850-853.

[8] Fedorenko G.A., Ivanova I.V., Sinyakov K.A. Improvement of the technological process of welding in the shielding gases in the wind. Welding production. 2010. 1. P. 6–13.

[9] Gribovsky G., Kravchuk B., Leninvkin V.A. Influence of the double annular flow of the shielding gases upon the process of consumable electrode welding. Welding production. 1996. 4. P. 6–8.

[10] Popravka D.L., Khvorostov N.E. Outdoor arc welding in the shielding gases. Mashinostroyeniye, 1979. 64 p.

[11] Ramsey G.M., Galloway A.M., Campbell S.W., McPherson N.A., Scanlon T.J. A computational fluid dynamic analysis of the effect of side draughts and nozzle diameter on shielding gas coverage during gas metal arc welding. (2012) Journal of Materials Processing Technology, 212 (8), pp. 1694-1699.

[12] Kang B.Y., Prasad Y.K.D.V., Kang M.J., Kim H.J., Kim I.S. Characteristics of alternate supply of shielding gases in aluminum GMA welding. (2009) Journal of Materials Processing Technology, 209 (10), pp. 4716-4721.

[13] Ley F.H., Campbell S.W., Galloway A.M., McPherson N.A. Effect of shielding gas parameters on weld metal thermal properties in gas metal arc welding. (2015) International Journal of Advanced Manufacturing Technology, 80 (5-8), pp. 1213-1221.
[14] Chinakhov D.A. Gas Dynamic Control of Properties of Welded Joints from High Strength Alloyed Steels. China Welding. 23. 3 (2014) pp. 27-31.
[15] Chinakhov D.A., Chinakhova E.D., Gotovschik Y.M., Grichin S.V. Influence of welding with two-jet gas shielding on the shaping of a welding joint. IOP Conf. Series: Materials Science and Engineering 125 (2016) 012013.

IOP Conf. Series: Materials Science and Engineering **221** (2017) 012003 doi:10.1088/1757-899X/221/1/012003

The Synthesis of Precise Rotating Machine Mathematical Model, Operating Natural Signals and Virtual Data

A A Zhilenkov and A A Kapitonov

ITMO University, 49, Kronverkskiy prospekt, Saint Petersburg, 197101, Russian Federation

E-mail: zhilenkovanton@gmail.com

Abstract. It is known that synchronous machines catalogue data are presented for the case of two-phase machine in rotating coordinate system, e.g. for their description with Park-Gorev's equation system. Nevertheless, many problems require control of phase currents and voltages, for instance, in modeling of the systems, in which synchronous generators supply powerful rectifiers. Modeling of complex systems with synchronous generators, semiconductor convertors and etc. (with phase currents control necessary for power switch commutation algorithms) becomes achievable with the equation system described in this article. Given model can be used in digital control systems with internal model. It doesn't require high capacity of computing resources and provides sufficient modeling accuracy.

1. Introduction

To analyze the processes in "synchronous generator – semiconductor converter" system it is necessary to have an adequate mathematical model. The simplifications made in the classic form of Park-Gorev's equations don't allow to derive this model, which will be shown further. Therefore, synchronous machines catalogue parameters are given right for the system describing those machines as two-phase machines in the rotating coordinate system.

Controlled rectifier (CR) and autonomous inverter (AI) are the ones considered nonlinear. Particular nonlinear properties of those elements are: relay nonlinearity; nonlinear parameters depend on time, which allows to classify them as the nonautonomous type; there are three and more nonlinearities in each CR and AI (for three-phase circuit).

Methods of analysis and synthesis of such systems are shown in the most accurate way in the nonlinear systems section in the theory of automatic control (TAC). With considering known nonlinear systems properties, the ones that lead to complicated equations, the amount of nonlinearities are usually limited by 1-2 [1, 2]. Despite lots of efforts to increase the accuracy of the calculations, it is noted, that "there are no universal analytical (mathematical) methods of calculating nonlinear systems" [3-4].

2. Problem statement

The synchronous generator (SG) model in Park-Gorev's equations (PGE) is applied in marine electric station models based on SG. This SG model has a lot of advantages when working on linear load, which are:

- minimal amount of SG differential equations is: 5 with only electric and 7 with electromechanical processes considered;

IOP Conf. Series: Materials Science and Engineering **221** (2017) 012003 doi:10.1088/1757-899X/221/1/012003

- models of SG voltage regulation due to excitation winding (EW) are the most simple, since PGE are reduced to EW;
- it is possible to get analytical equations (with some uncomplicated assumptions) for SG short-circuit mode, SG output voltage, etc.

Modeling of SG's work on rectifier load with PGE doesn't give good results for the following reasons:

- each stator phase winding is connected with its nonlinearity formed by controlled rectifier (CR) phase circuit. Reducing equations of stator three windings to the armature can be accomplished in 2 steps: first is using linear phase conversion (of three-phase stator winding signals in ABC axes into the two-phase stator winding signals in fixed α-β axes) and then nonlinear coordinates conversion (of two-phase stator winding signals in fixed α-β axes into the two-phase stator winding signals in d-q axes). This reducing also converts 3-6 CR nonlinearities (3 for half-bridge rectifier and 6 for full-bridge rectifier) into two huge nonlinearities in stator windings circuits in d-q axes;
- in SG model there will be 3-6 firstly nonlinear elements, originated from CR, and also nonlinearities of coordinate conversions from α-β into d-q;
- nonlinear elements (3 or 6) go through conversions, which represent the analytical form of structural conversions of nonlinear elements joints in the system;
- converted nonlinear joints' input and output signals don't equal the ones defining CR's thyristor's VS (diode's) state, - voltage u_{VS}, current i_{VS}.

From the stated reasons it is clear, that the modeling of SG, working on CR, in PGE leads to the sphere of problems, which are not only unsolved in TAC, but also considered best not be addressed at all. Therefore, in the electric station model working on semiconductor converter, it is justifiable to neglect SG model in PGE in favor of a model with SG stator windings natural phase signals instead – voltages u_A, u_B, u_C and currents i_A, i_B, i_C.

3. Conversion of SG model

The scheme of loaded SG with natural windings on stator and without neutral point output is shown on fig.1 [5].

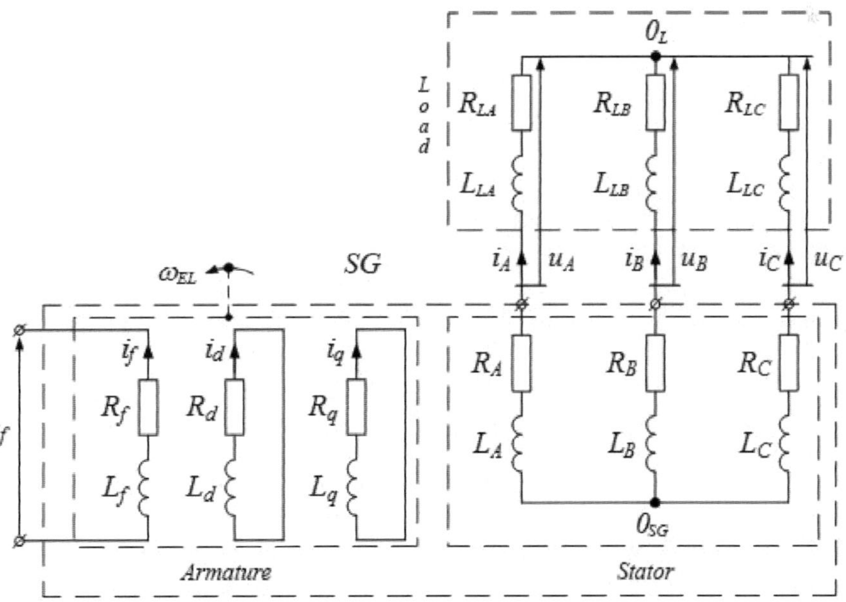

Figure1. The scheme of SG for three-wire circuit with physical stator and armature windings

On this scheme in all SG windings there are physical currents i_A, i_B, i_C, i_f, i_d and i_q, armature rotates relative to stator with frequency ω_{EL}. SG for the given scheme is described in the following system of equations:

$$\begin{cases} i_A + i_B + i_C = 0, \\ 0 = (R_A + R_{нA})i_A + p\Psi_A, \\ 0 = (R_B + R_{нB})i_B + p\Psi_B, \\ 0 = (R_C + R_{нC})i_C + p\Psi_C, \\ u_f = R_f \cdot i_f + p\Psi_f, \\ 0 = R_d \cdot i_d + p\Psi_d, \\ 0 = R_q \cdot i_q + p\Psi_q, \\ \Psi_A = (L_A + L_{нA})i_A + L_{AB}i_B + L_{AC}i_C + L_{Af}i_f + L_{Ad}i_d + L_{Aq}i_q, \\ \Psi_B = L_{AB}i_A + (L_B + L_{нB})i_B + M_{BC}i_C + L_{Bf}i_f + L_{Bd}i_d + L_{Bq}i_q, \\ \Psi_C = L_{AC}i_A + L_{BC}i_B + (L_C + L_{нC})i_C + L_{Cf}i_f + L_{Cd}i_d + L_{Cq}i_q, \\ \Psi_f = L_{Af}i_A + L_{Bf}i_B + L_{Cf}i_C + L_f i_f + L_{fd}i_d + L_{fq}i_q, \\ \Psi_d = L_{Ad}i_A + L_{Bd}i_B + L_{Cd}i_C + L_{fd}i_f + L_d i_d + L_{dq}i_q, \\ \Psi_q = L_{Aq}i_A + L_{Bq}i_B + L_{Cq}i_C + L_{fq}i_f + L_{dq}i_d + L_q i_q. \end{cases} \quad (1)$$

System (1), which consists of 12 equations (without consideration of generator's mechanical part), contains one independent variable – excitation voltage u_f and 12 state variables – currents $i_A, i_B, i_C, i_f, i_d, i_q$ and linkages $\Psi_A, \Psi_B, \Psi_C, \Psi_f, \Psi_d, \Psi_q$. Therefore, the necessary condition of system's solvability is satisfied.

When calculation of phase currents i_A, i_B, i_C is done, phase voltages can be found from the equations

$$u_A = (pL_{нA} + R_{нA})\, i_A .$$

However, self-inductances (with one-letter indexes) and mutual inductances (with two-letter indexes) L in the equations (1) are periodical coefficients of rotation angle φ_{EL} of armature relative to stator. There is no information on L instant values in SG catalogue database. It's commonly accepted to place data concerning L inductances suitable for SG model in PGE into the database, and for this model the inductances have constant values [6]. The property of L constancy in the PGE model is achieved through the fact, that all SG windings, when reduced to the excitation winding, are motionless relative to each other.

Let's make a system of equations describing SG as a two-phased machine in random axes in u–v coordinates, which are common for all SG windings (both armature and stator) and rotating in space with random velocity ω_K [7].

The result of it is a system of equations describing SG as two-phase machine in random axes u-v, rotating with velocity ω_K. With any value of velocity ω_K the inductances L and mutual inductances M, through which $\Psi_{1u}, \Psi_{1v}, \Psi_{fu}, \Psi_{fv}, \Psi_{2u}$ are calculated, will have constant values, since all SG model windings are motionless relative to each other. Moreover, and this is very important, with any choice of coordinate axes u–v and their frequency $\omega_K L$ values do not change, only voltage frequencies ($u_{1u}, u_{1v}, u_{fu}, u_{fv}$) do.

Let's consider coordinate axes u–v incorporated with axes d-q of the excitation winding. Then their rotation frequency ω_K becomes close to the armature frequency ω_{EL} (fig.1) [8-9]. After placing $\omega_K = \omega_{EL}$ into the system (2) and changing of indexes $u \to d$ and $v \to q$ we derive

$$\begin{cases} u_{1d} = R_1 \cdot i_{1d} + p\Psi_{1d} - \omega_{EL} \cdot \Psi_{1q} \\ u_{1q} = R_1 \cdot i_{1q} + p\Psi_{1q} + \omega_{EL} \cdot \Psi_{1d} \\ u_f = R_f \cdot i_f + p\Psi_f \\ 0 = R_2 \cdot i_{2d} + p\Psi_{2d} \\ 0 = R_2 \cdot i_{2q} + p\Psi_{2q} \end{cases}, \quad (2)$$

IOP Conf. Series: Materials Science and Engineering **221** (2017) 012003 doi:10.1088/1757-899X/221/1/012003

in which the equation for the excitation winding on the q axis is neglected, since there is literally no such winding, and index d is neglected in the equation for the excitation winding on the d axis.

System (2) is a PGE system, in which SG catalogue data can be used for calculation of the linkages.

Let's consider [10-12] that coordinate axes u–v coincide with axes α-β, also axis α coincides with axis A. Therefore, rotation frequency ω_K of coordinate axes α-β equals zero. Now the SG model contains fictitious voltages $u_{1\alpha,}u_{1\beta}$ and currents $i_{1\alpha,}i_{1\beta}$ of stator with two windings in α-β axes. For modeling of semiconductor converter, working due to SG, it is necessary to operate real voltages $u_{A,}u_{B,}u_C$ and currents $i_{A,}i_{B,}i_C$ in stator circuit. Since that, it is necessary to convert system, which describes two-phase machine, into three-phase machine through the stator. With conversion of a two-phase machine into three-phase we solve a problem of conversion variables of equivalent two-phase machine stator in coordinate system α-β into the real variables of three-phase machine stator in axes A-B-C.

The formulas of direct conversion of currents $i_{1\alpha}, i_{1\beta}$ into currents i_A, i_B, i_C, which originate from the condition of invariability of SG power, calculated in axes α-β and A-B-C, are viewed as [4,11]. Let's change currents $i_{1\alpha}, i_{1\beta}$ and voltages $u_{1\alpha}, u_{1\beta}$ (in α-β axes) on currents i_A, i_B, i_C and voltages u_A, u_B, u_C (in axes A-B-C) in equations of the system and linkages equations. Considering mechanical part, we derive the following system of SG equations:

$$\begin{cases} u_A = R_1 i_A + p\Psi_A \\ u_B = R_1 i_{1B} + p\Psi_B \\ u_C = R_1 i_{1C} + p\Psi_C \\ U_{2\alpha} = R_2 i_{2\alpha} + p\Psi_{2\alpha} + \omega_{EL}\Psi_{2\beta} \\ U_{2\beta} = R_2 i_{2\beta} + p\Psi_{2\beta} - \omega_{EL}\Psi_{2\alpha} \\ U_{f\alpha} = R_f i_{f\alpha} + p\Psi_{f\alpha} + \omega_{EL}\Psi_{f\beta} \\ U_{f\beta} = R_f i_{f\beta} + p\Psi_{f\beta} - \omega_{EL}\Psi_{f\alpha} \end{cases} \quad (3)$$

and linkages equations with mechanical part:

$$\begin{cases} \Psi_A = L_1 i_{1A} + \sqrt{\frac{2}{3}} L_{12}(i_{2\alpha} + i_{f\alpha}) \\ \Psi_B = L_1 i_{1B} + \frac{1}{\sqrt{2}} L_{12}(i_{2\beta} + i_{f\beta}) - \frac{1}{\sqrt{6}} L_{12}(i_{2\alpha} + i_{f\alpha}) \\ \Psi_C = L_1 i_{1C} - \frac{1}{\sqrt{2}} L_{12}(i_{2\beta} + i_{f\beta}) - \frac{1}{\sqrt{6}} L_{12}(i_{2\alpha} + i_{f\alpha}) \\ \Psi_{2\alpha} = L_2 i_{2\alpha} + L_{12}\left(\sqrt{\frac{3}{2}} \cdot i_A + i_{f\alpha}\right) \\ \Psi_{2\beta} = L_2 i_{2\beta} + L_{12}\left(\frac{1}{\sqrt{2}} \cdot (i_B - i_C) + i_{f\beta}\right) \\ \Psi_{f\alpha} = L_f i_{f\alpha} + L_{12}\left(i_{2\alpha} + \sqrt{\frac{3}{2}} \cdot i_A\right) \\ \Psi_{f\beta} = L_f i_{f\beta} + L_{12}\left(i_{2\beta} + \frac{1}{\sqrt{2}} \cdot (i_B - i_C)\right) \\ \Delta\omega(t) = \frac{1}{2J}\int_0^t (M_{MEX} - M_{EL})dt \\ \omega_{EL}(t) = \Delta\omega(t) + \omega_S \end{cases} \quad (4)$$

where $\Delta\omega(t)$ is a difference between armature rotation speed and synchronous speed; J is armature moment of inertia; M_M is mechanical moment; M_{EL} is electrical moment; $\omega_{EL}(t)$ is armature rotation speed; ω_S is synchronous speed.

Conclusion

We derived a full system of equations describing synchronous generator in still axes with the usage of catalogue parameters and operating real currents and voltages of stator circuit (three-phase machine).

This model can be applied in digital control systems with internal model, giving advantage on information processing speed and modeling accuracy, which will be shown in further publications.

Acknowledgments

This work was supported by the Government of the Russian Federation, Grant 074-U01. This work was supported by the Ministry of Education and Science of Russian Federation (Project 14.Z50.31.0031).

Reference

[1] Chernyi S and Zhilenkov A 2015 *Transport and Telecommunication.* Modeling of Complex Structures for the Ship's Power Complex Using Xilinx System **16** 73-82

[2] Zhilenkov A and Efremov A 2017*IOP Conference Series: Materials Science and Engineering*A quality evaluation of stabilization of rotation frequency of gas-diesel engines when using an adaptive automatic control system **177** No. 1 012043

[3] Zhilenkov A and Efremov A 2017 *IOP Conference Series: Earth and Environmental Science*Evaluation Of Rotation Frequency Gas-Diesel Engines When Using Automatic Control System **50** 1-7

[4] Gillan M J 1991 *Computer Simulation in Materials Science* ed M Meyer and V Pontikis (Dordrecht: Kluwer) p 257

[5] Zhilenkov A 2016 *Vibroengineering PROCEDIA.* The study of the process of the development of marine robotics **8**

[6] Stich I, Payne M C, King-Smith R D, Lin J-S and Clarke L J 1992 *Phys. Rev. Lett.* **68** 1351

[7] Nicolet C, Greiveldinger B, Herou J, Kawkabani B, Allenbach P, Simond J and Avellan F 2007 High-order modeling of hydraulic power plant in islanded power network *IEEE Trans. Power Syst.***22** 1870-1880 Nov.

[8] Pannatier Y., Kawkabani B., Nicolet C., Simond J. J., Schwery A. and Allenbach P. 2010 Investigation of control strategies for variable-speed pump-turbine units by using a simplified model of the converters *IEEE Trans. Ind. Electron.***57** 3039-3049

[9] Gorbatov V A, Smirnov M I and Khlytchev I S 1991 (Moscow: Energoizdat)

[10] Kundur P, Balu N and Lauby M 1994 Power system stability and control 1176

[11]Padoan A. C., Kawkabani B., Schwery A., Ramirez C., Nicolet C., Simond J. J. and Avellan F. 2010 Dynamical behavior comparison between variable speed and synchronous machines with PSS *IEEE Trans. Power Syst.* **25** 1555-1565

[12] Mitrovic N, Kostic V, Petronijevic M and Jeftenic B 2009 *Advances in Electrical and Computer Engineering* **9** 57-62

IOP Conf. Series: Materials Science and Engineering **221** (2017) 012004 doi:10.1088/1757-899X/221/1/012004

Simulink-aided Design and Implementation of Sensorless BLDC Motor Digital Control System

A A Zhilenkov, Y N Tsvetkov, V B Chistov, A P Nyrkov and S S Sokolov [a]

Admiral Makarov State University of Maritime and Inland Shipping, 5/7, Dvinskaya str., St. Petersburg,198035, Russia

E mail: [a] sokolovss@gumrf.ru

Abstract. The paper describes the process of creating of brushless direct current motor's digital control system. The target motor has no speed sensor, so back-EMF method is used for commutation control. Authors show how to model the control system in MatLab/Simulink and to test it onboard STM32F4 microcontroller.This technology allows to create the most flexible system, which will control possible with a personal computer by communication lines. It is possible to examine the signals in the circuit of the actuator without any external measuring instruments – testers, oscilloscopes, etc. – and output waveforms and measured values of signals directly on the host PC.

1. Introduction

Brushless direct current motor (BLDC) is becoming more widely used in various technical systems. From warships to tiny drones. In addition to the known advantages of these motors they have one major drawback – the need to control the rotor position. In most cases, their design does not include position sensors, which increases their reliability, and simplifies design, but it complicates the management process.

The study of the principles of management of such motors is necessary when training. Therefore, we have developed a number of laboratory applications including BLDC motor and control system. The installation can be programmed and controlled from MATLAB/Simulink. They also enable research and testing of adaptive control systems.

2. The structural design of the laboratory bench

The laboratory bench includes a control system, a control object, the set of feedback to monitor the state of the control object, the communication line with a PC for programming and debug the program loaded in the control system [1], as well as a line of communication with the personal computer to enable communication of the control system with control software and displays, installed on a personal computer.

This structure allows to create the most flexible system, which will control possible with a personal computer by communication lines. It is possible to examine the signals in the circuit of the actuator without any external measuring instruments – testers, oscilloscopes, etc. – and output waveforms and measured values of signals directly on the PC screen.

The structuraldiagram shown in figure 1.

Programming and debugging are loaded in the DSPS programs is made through a separate line ST-Link v2. Data exchange with a personal computer through the USB port of DSP.

The currents of the windings U, V and W are commutated by CMOS (or MOSFET) transistors [2] of the inverter, the channels of which are controlled by the signals Ch1-3 and Ch1n-3n of TIM8 timer counters.

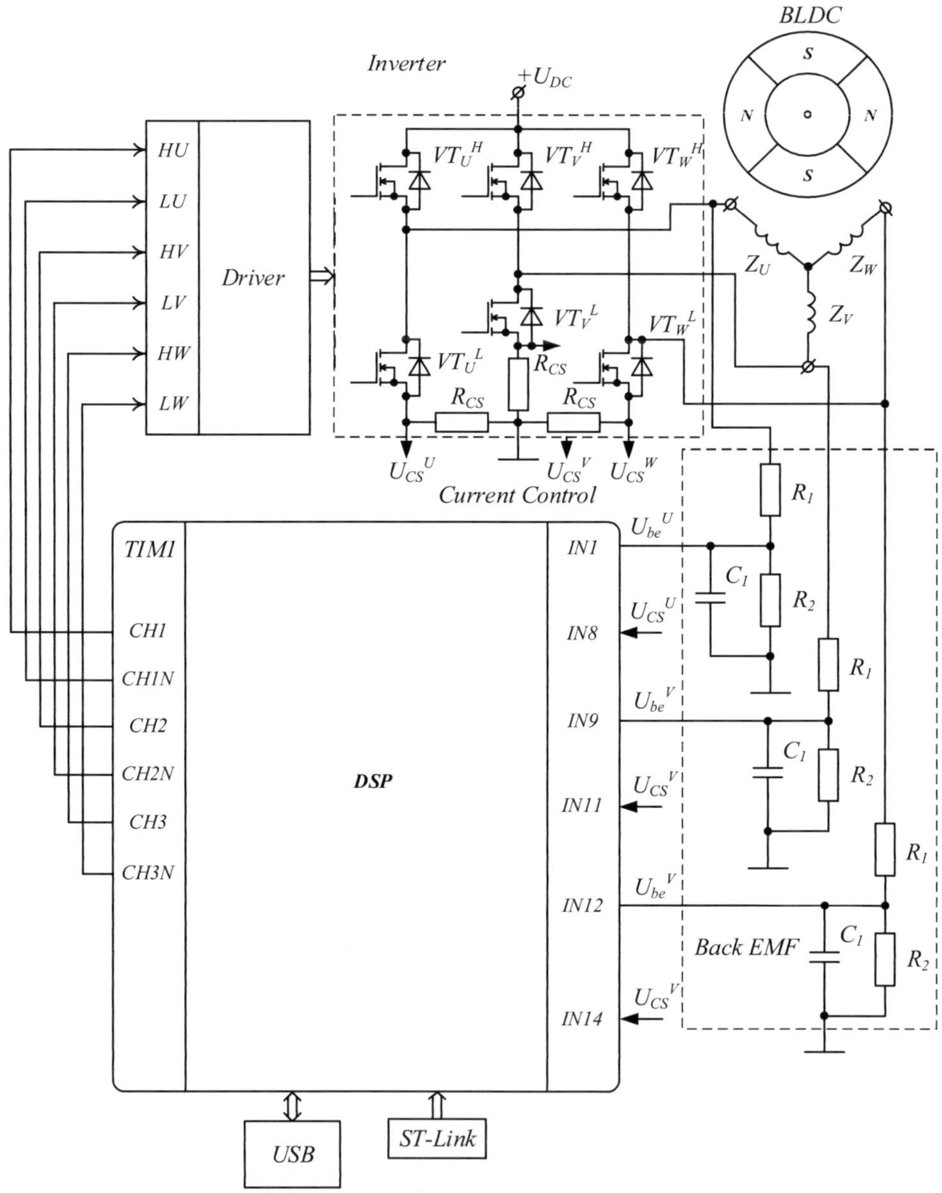

Figure1. The functional diagram design of the laboratory bench.

The position control of the BLDC's rotor is implemented using monitor value of the back EMF in (at the moment) the coil, one end of which is currently "in the air", i.e. not connected to ground nor to "+" supply. At this time, the winding rotating field of the rotor induced EMF the signal of which it is possible to determine the time of switching of the winding. Back EMF signals arrive at the inputs of the ADC unit.

Monitoring the current in each winding during the opening of the relevant transistor takes place via measurement of the voltage drop on the source resistor Rcs.

3. ARM Cortex-M microcontroller based control system

Modeling and writing a program control system is made in MATLAB/Simulink [3]. The program consists of two main blocks – start BLDC and the maximum speed with the office of the induced EMF in the windings.

Thus, the identification of the position of the rotor, due to the lack of built-in plug-in position sensors of the rotor is performed by controlling the induced EMF in the free phase. In each moment of time one of the phases of the motor is connected "+" power supply, and the other is "-", and the remaining third phase is not connected to the power source, as if hanging in the air. Rotating the rotor moves through installed on the powerful permanent magnets of EDS in this free winding. As a result, change of magnetic poles extending in the rotation by the loose winding, the voltage on it is changed by changing the induced EMF. Measuring voltage on the free phase, we can determine the time of switching to the next position of the rotor. Usually determine the time of the voltage transition to the free phase through the zero point (half of the supply voltage). That is, system needs to track the moment when the passive phase's voltage will be equal to the average point. Of course, for this method to work, the motor must rotate. This method works well at relatively high motor speeds. At low speed induced EMF may be insufficient to clearly determine the rotor position. However, this method is applied.

However, in most cases the midpoint is not available. I.e. there is no way to physically connect to it without disassembling the motor. In this case, often use the scheme with the creation of a virtual central point.This scheme is very simple and is used very often [4], but it has its-drawbacks. For speed control BLDC motors use PWM to regulate power accordingly. Due to the PWM voltage mid-point is not permanent. It varies in a wide range of voltages. In order for the microcontroller was able to measure the voltage apply circuit coordination of the signals, voltage dividers and RC filters to smooth out fluctuations (Fig. 1).Thus, the midpoint voltage can be in the range from 0 to the supply voltage. In this case, for matching the signal levels of the digital control system (DCS) the necessary additional circuit solutions. In addition, to prevent accidents it is necessary to apply protection schemes the DCS.The presence of the voltage divider leads to a decrease in sensitivity at low motor speeds, and the presence of filter introduces latency.The delay is causing errors in determining the position of the rotor and may be critical for motor control at high speeds.Since the circuit with the virtual mid-point quite "noisy", it is possible to resort to another scheme where stable voltage mid-point is set separately and is dependent on the supply voltage.

Consider implementing the control of BLDC motors with digital control system.For formation of desired forms of voltages in phases and plug-phase displacements between the voltages necessary to implement a particular algorithm of switching of power switches of the inverter plug. For BLCD motors with trapezoidal distribution of magnetic field shape of the voltages in the windings shall be as given in figure 2 [4-5].

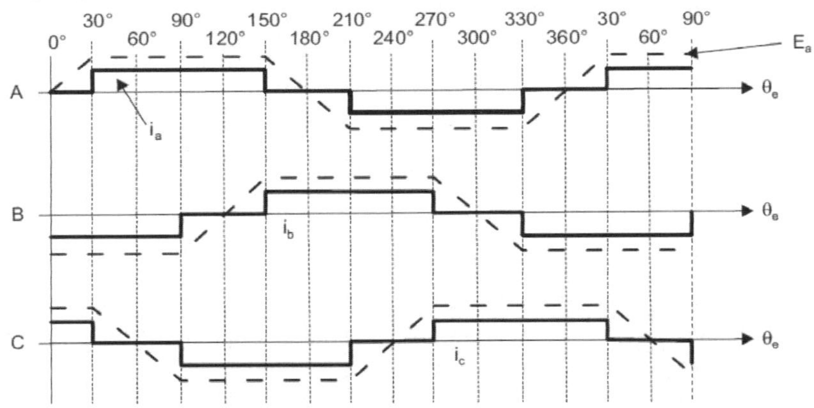

Figure 2. The switching sequence for all phases.

As can be seen from the diagrams, the phase is formed of a three-phase sequence voltages are shifted relative to each other at angle 120^0. The data voltage can be generated switching sequence described by the truth table.This switching sequence has been implemented in the "Table of Truth" block. The internal structure of this block is presented in figure 3.

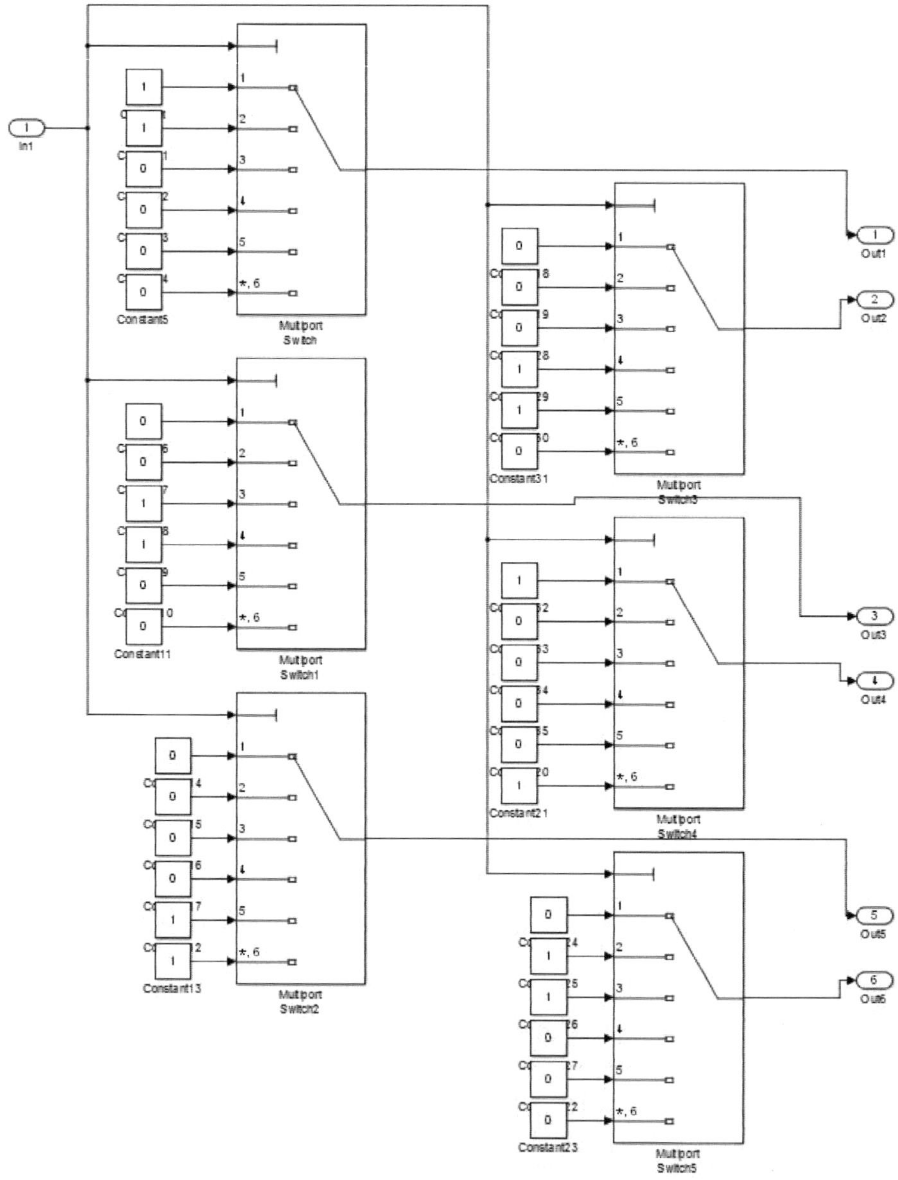

Figure 3. The "Table of Truth" block implementation in Simulink.

The input "in1" of the block is fed a digital code represented by the number varying from 1 to 6, which, as a multiband switches (Multiport Switch) depending on the values of the received code commute from one of six inputs of each switch on its output. Six outputs of the six switches correspond to the control channels of the respective power switches of the inverter plug.

The code input unit is formed by a counter (Fig. 4). Before applying for the entrance block of the truth table to the output code of the counter adds a unit to code the input to satisfy the input combinations of the truth table.

IOP Conf. Series: Materials Science and Engineering **221** (2017) 012004 doi:10.1088/1757-899X/221/1/012004

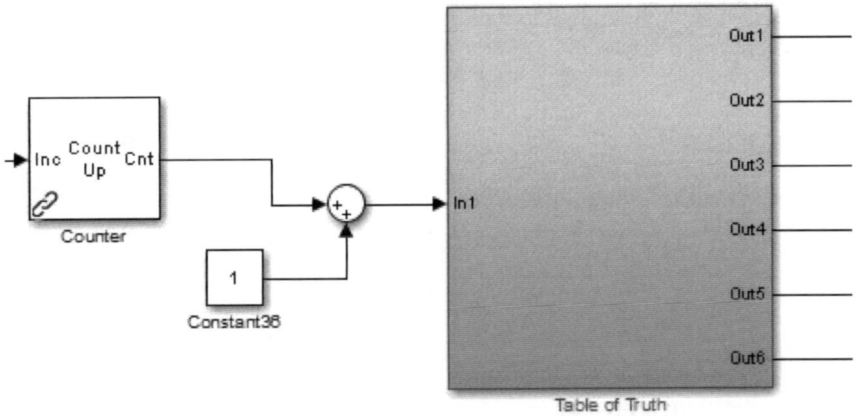

Figure 4. The implementation of the switching sequence in the block.

The pulse input of the counter are supplied either from the internal setpoint at the input “Clock”, or from the circuit of detection of the moments of switching of the induced EMF. The source select signal is produced via the control input of the “Switch” block (Fig. 5).

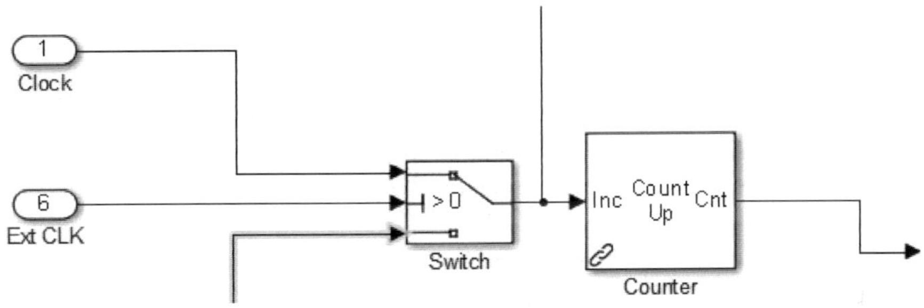

Figure 5. The formation of the numerical sequence input truth table.

4. Detection of the moments of switching of the induced EMF

The detection scheme presented in the corresponding drawing graphic part. The scheme detects the achievement of the induced EMF thresholds “REFp” or ”REFn”, corresponding to transitions of the EMF from the negative value to the positive and vice versa [5,13,14].

As noted, this method of detection does not work until the rotor is stationary. Therefore, the motorstart is carried out with supply on the winding of the sequence of voltages of low frequency. Created by the rotary field at the end of the poles "interlock" with the magnetic field of the rotor and drags it along, bringing the rotor from the stationary state. As a rule, before the rotor comes in random movement, the twitching of the rotor, often oscillatory in nature, which can be seen in the diagrams of signals of the drive.

When the rotor comes in uniform rotation, its speed can increase accelerating the change of condition of power switches according to the truth table. And then it can go to the control mode according to the detection.In this mode, the time of change of the output status table will be determined not by hard account the above-described counter, namely identifying moments.

As shown by case studies on a laboratory bench(Fig.6, 7), form, character transitions and thresholds changes with the rotational speed of the rotor and load.From figure 7 it is seen that the uncorrected thresholds that gave an accurate detection at a low speed of rotation of the BLDC (Fig. 6), starts to give error detection with increasing speed of rotation.

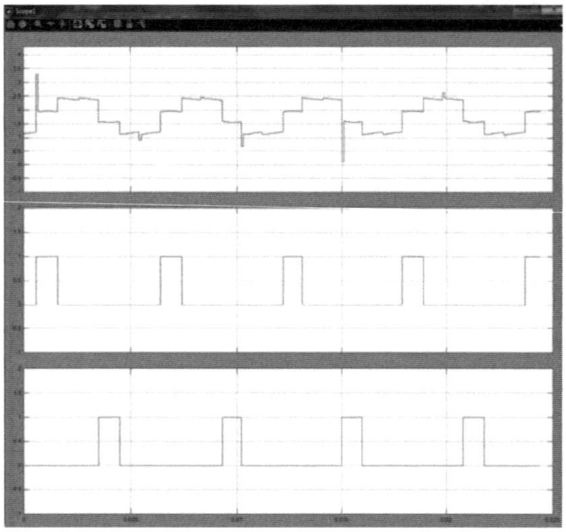

Figure 6. Selected for a given rotation speed threshold values give an accurate detection of the moments of switching.

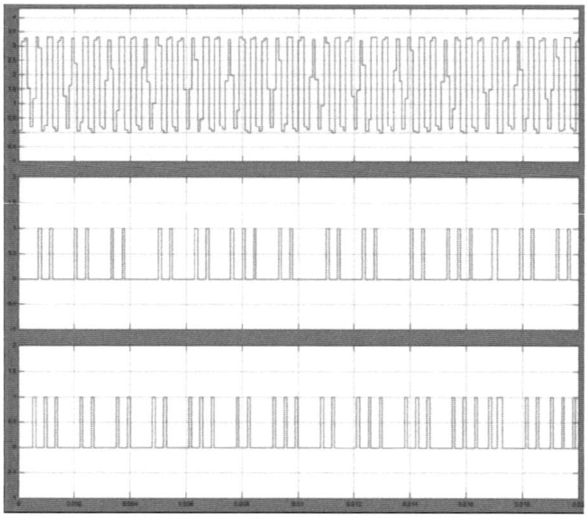

Figure 7. Selected threshold values give errors when detecting the moments of switching.

In this regard, the control program introduced the ability to set thresholds in real-time through the communication control program on a PC with a digital control system.As can be seen, with the change of parameters of rotation of the motort, its load, the shape of the induced EMF, in many cases radically different from the theoretical shape. All this significantly complicates the process of automatic control of the BLDC motors without rotor position sensor, which requires a search for appropriate technical solutions.

The laboratory equipment is shown in Figure 8.

Figure 8. The laboratory equipment.

5. Conclusion

We have created the laboratory bench which allows us to study the BLDC motor and control systems of modern polyphasemotors. The system allows to simulate and control all the main signals in the power circuits and the control system by means of MATLAB/Simulink environment.

The system can be used for educational and research purposes.

IOP Conf. Series: Materials Science and Engineering **221** (2017) 012004 doi:10.1088/1757-899X/221/1/012004

6. Reference

[1] Chernyi S and Zhilenkov A 2015 *Transport and Telecommunication.* Modeling of Complex Structures for the Ship's Power Complex Using Xilinx System **16** 73-82

[2] Zhilenkov A and Efremov A 2017*IOP Conference Series: Materials Science and Engineering*A quality evaluation of stabilization of rotation frequency of gas-diesel engines when using an adaptive automatic control system **177** No. 1 012043

[3] Zhilenkov A and Efremov A 2017 *IOP Conference Series: Earth and Environmental Science*Evaluation Of Rotation Frequency Gas-Diesel Engines When Using Automatic Control System **50** 1-7

[4] Gillan M J 1991 *Computer Simulation in Materials Science* ed M Meyer and V Pontikis (Dordrecht: Kluwer) p 257

[5] Zhilenkov A 2016 *Vibroengineering PROCEDIA.* The study of the process of the development of marine robotics **8**

[6] Stich I, Payne M C, King-Smith R D, Lin J-S and Clarke L J 1992 *Phys. Rev. Lett.* **68** 1351

[7] Nicolet C, Greiveldinger B, Herou J, Kawkabani B, Allenbach P, Simond J and Avellan F 2007 High-order modeling of hydraulic power plant in islanded power network *IEEE Trans. Power Syst.***22** 1870-1880 Nov.

[8] Pannatier Y., Kawkabani B., Nicolet C., Simond J. J., Schwery A. and Allenbach P. 2010 Investigation of control strategies for variable-speed pump-turbine units by using a simplified model of the converters *IEEE Trans. Ind. Electron.***57** 3039-3049

[9] Gorbatov V A, Smirnov M I and Khlytchev I S 1991 (Moscow: Energoizdat)

[10] Kundur P, Balu N and Lauby M 1994 Power system stability and control 1176

[11]Padoan A. C., Kawkabani B., Schwery A., Ramirez C., Nicolet C., Simond J. J. and Avellan F. 2010 Dynamical behavior comparison between variable speed and synchronous machines with PSS *IEEE Trans. Power Syst.* **25** 1555-1565

[12] Mitrovic N, Kostic V, Petronijevic M and Jeftenic B 2009 *Advances in Electrical and Computer Engineering* **9** 57-62

[13] Budnik V, Chernyi S 2016 *Procedia Engineering* **150** 2150-2156 http://dx.doi.org/10.1016/j.proeng.2016.07.256

[14] Chernyi S 2016 *IOP Conference Series: Materials Science and Engineering* **124** 012048.

IOP Conf. Series: Materials Science and Engineering **221** (2017) 012005 doi:10.1088/1757-899X/221/1/012005

Development of Cross-Platform Software for Well Logging Data Visualization

R K Akhmadulin [a], **A I Miraev**

Industrial University of Tyumen, 38 Volodarskogo st., Tyumen, 625000, Russia

E-mail: [a] ark@tsogu.ru

Abstract: Well logging data processing is one of the main sources of information in the oil-gas field analysis and is of great importance in the process of its development and operation. Therefore, it is important to have the software which would accurately and clearly provide the user with processed data in the form of well logs. In this work, there have been developed a software product which not only has the basic functionality for this task (loading data from .las files, well log curves display, etc.), but can be run in different operating systems and on different devices. In the article a subject field analysis and task formulation have been performed, and the software design stage has been considered. At the end of the work the resulting software product interface has been described.

1. Introduction

Processing and view of data acquired during well logging plays an important role in oil-gas production. The acquired data are essential in the development simulation model analysis and design in general as well as in the development of alternative oil production technologies [1, 2]. The information about well logging can be represented in digital form and in files of .las format.

A powerful tool in this process is the use of specialized software which can accurately and clearly provide the user with processed data in the form of well logs. Depending on a level of automation, such software can allow to solve one of the tasks of the well log analysis –identification of reservoirs containing oil, gas and water fluids. Theoretically, such software can be used under different operating systems and on different platforms [3].

This work covers the development of software for file parsing in the .las format and provision the user with necessary information about resulting well logging in the form of well log curves.

2. Research

The creation of cross-platform software for well logging visualization is an important task. It is stipulated by two factors. First, the developed software allows well logging data processing. Secondly, this product is cross-platform, which is quite efficient and profitable at the moment. A distinctive feature of this software product, apart from its cross-platform nature, is its ability to connect new plug-ins that can extend the functional of the product in accordance with the user's requirements.

*.las files analysis when loading implies automatic recognition and correct downloading files of different versions. Depending on which file version has been downloaded the software will run that parser which corresponds to the current file structure. This frees the user from binding to a definite *.las file specification.

Before downloading the data files into the program their structure was analyzed. *.las files have three different standards: version 1.2, version 2.0 and version 3.0. These standards have been worked out by Canadian Well Logging Society. All the three standards have a common structure which contains four main sections: "~Version Information", "~Well Information", "~Curve Information" and "~ASCII Log Data" (Fig. 1).

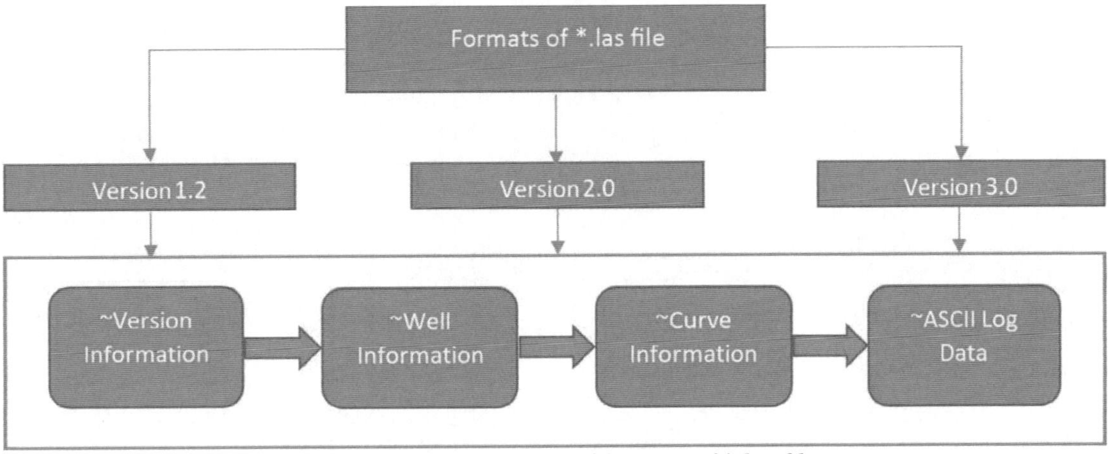

*Figure 1 – The structure of formats of *.las files*

The format .las version 2.0 file has been developed as a result of correcting the most common errors encountered in the format .las version 1.2 file. The .las version 3.0 file is considered the newest format, which incorporated all the required data and was created with account of the experience of versions 1.2 and 2.0. In Russia different versions of .las files are widespread.

The file section, which is read first, is "~Version Information". It contains the information about the file version. After the definition of the file version it is necessary to read the information about the well in the section of "~Well Information". This section is different for versions 1.2 and 2.0. The next section is "~Curve Information" which contains the information about the curves which in its turn includes the names of the applied techniques and their units of measurement respectively. Each well logging technique is provided with comments. This section is similar in version 1.2 and 2.0 files but has differences in version 3.0 files. The last section, that of "the~ASCII Log Data", contains a dataset in the numerical form according to the techniques described in the section of "~Curve Information". This section is also absolutely similar in versions 1.2 and 2.0, but differs in version 3.0.

The results of the files structure analysis in the .las format have become the basis of the parser class diagram. The diagram is shown in Fig. 2.

The parser architecture includes the class of "ReadersLauncher" which is the main class that reads from a file and then transmits the data stream. The classes of "LasReaderVersion" alternately read a section with a version until the class corresponding to the file version switches on. Next goes reading of all the sections and data recording in the corresponding data containers. All the data containers are based on the abstract class of "KeyWDataBase". As a result of reading data from a file, all the read sections are transferred to the class of "LasData", and later to the class of "ReadersLauncher".

Non-functional requirements apart from functional ones have become an important component at the design stage of a real software product. So, the main ones are as follows:

- graphics, pleasurable for the user, and the possibility to work on the average personal computer without a high load on it;
- the support of three popular operating systems: Windows, Linux, and Mac.

Elaborate class architecture with clear partitioning of responsibility was to be designed for the software. This can be achieved with the help of object-oriented programming using the programming pattern Model-View-ViewModel (MVVM). One of the advantages of this approach is simple extendibility and the possibility of connecting plugins with the main program.

IOP Conf. Series: Materials Science and Engineering **221** (2017) 012005 doi:10.1088/1757-899X/221/1/012005

The final result of well logging data processing in the software is visualization of well logs for their subsequent analysis by the user. Additional modules (plugins) allow displaying tops and bottoms of layers as well as levels of oil and water occurrence.

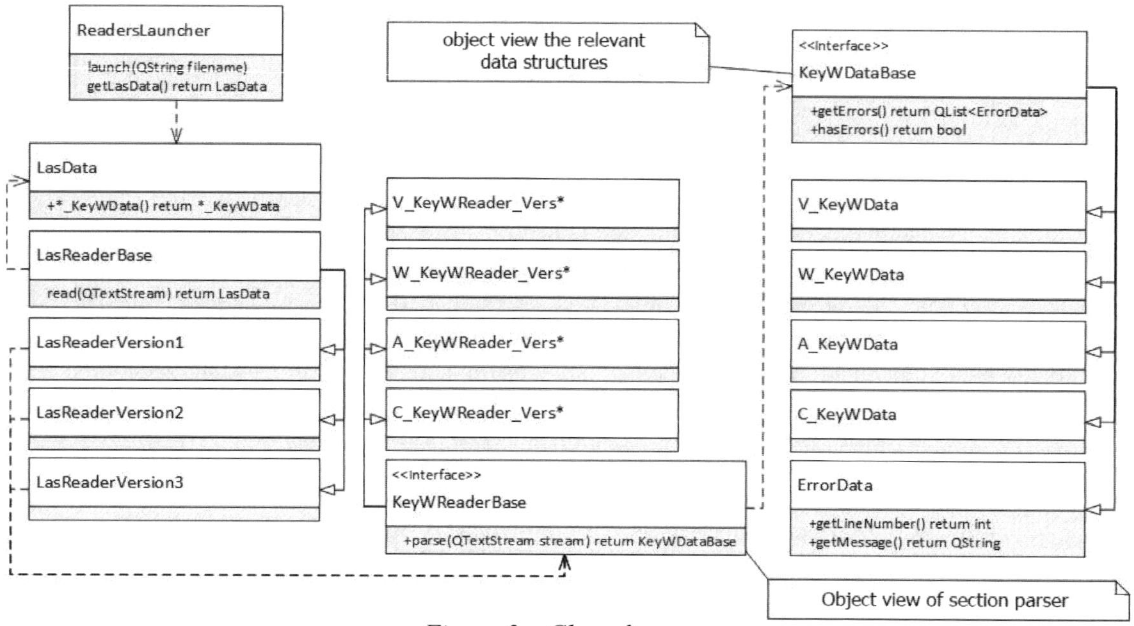

Figure 2 – Class diagram

Currently, when conducting geophysical surveys different-purpose software is widely used [4-6]. However, the known software products with similar functionality typically represent software systems with abundance of different functions (including the ones not required to solve the considered in the present work task) and work on a single platform – Windows. Thus, there are no analogues software with the functional similar to that considered in this work, able to operate on various platforms.

The software product design was carried out using the entity-relationship methodology to design back-end database and standardized language for object-oriented modeling UML. Use of object-oriented approach is one of the basic principles of design and has wide application in solving such problems [7-10]. Further work on the development of the program was carried out in accordance with the built at this stage models.

Since the program is designed to work with large volumes of data, it was very important to select a programming language which would be high performance. Such qualities are intrinsic to the family of programming languages C. To accomplish this work, according to the results of experiments, the C++ language was chosen. This language is widely used for the development of the various-purpose systems, including those based on complicated mathematical models and using large amounts of data, such as [11-14].

As a development environment it was necessary to choose a tool which would allow the development of a cross-platform product. In the end, it was Qt Creator – a cross-platform and open-source software development environment in QML, C++ and C.

It should be noted that Qt Creator is provided with a UI development technology Qt Quick, which allows creating your own interface in a declarative QML with minimal effort. There is also a technology of signal slots, which allows dividing the user interface and program logic into two components independent from each other, which adds flexibility in design and further modification of software.

PostgreSQL has been chosen as a database management system. This is an open-source system for database management that allows downloading unlimited amounts of data with high speed and easy-to-understand interface.

The development of the software was carried out in two stages: core development and creation of the user interface.

The core was developed with consideration of the fact that software components should be independent from each other to the maximum to ensure further extendibility without redesigning its architecture. Programming language C++ allowed achieving this independence; and the performance of the system showed a high speed under testing conditions.

The GUI was developed on the basis of QML using the technology of data transfer between a view and view model by means of signals-slots. As a result, an interface, almost independent from the software core was created, which corresponds to the task of MVVM pattern implementation.

Immediately after the main program was created the functional was extended in the form of a plug-in to display the boundaries of the tops and bottoms of layers and to display the areas of water and oil occurrence. Such data will provide a specialist with more information for further analysis of well logging data (Fig. 3). Data are downloaded from format *.csv files.

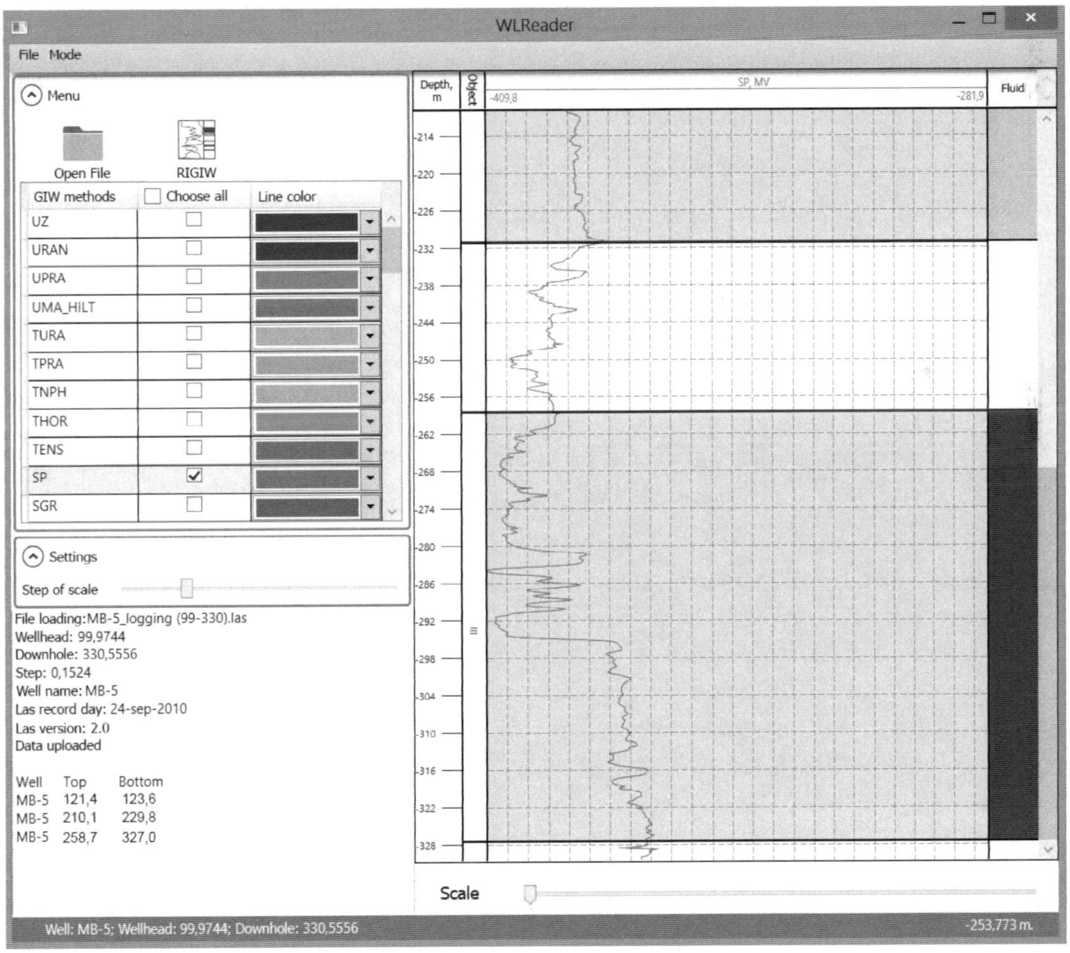

Figure 3 – The program with the connected plugin

IOP Conf. Series: Materials Science and Engineering **221** (2017) 012005 doi:10.1088/1757-899X/221/1/012005

As a result, the developed software has a clear interface and fully performs its task of well logging data visualization. The program can work under various operating systems without changing the initial interface and logic. The software was tested in Windows, Linux and Mac, and able to work on mobile devices, tablets, and other platforms. The developed product fully meets all the requirements, namely:

- visualizes well logging data;
- works under different platforms;
- allows to connect additional plugins.

After reading the *.las file all the well logging techniques are displayed in the main window (Fig. 4) in the form of a list. The possibility of selecting curves to be displayed on the screen has been implemented. Scaling data can be performed.

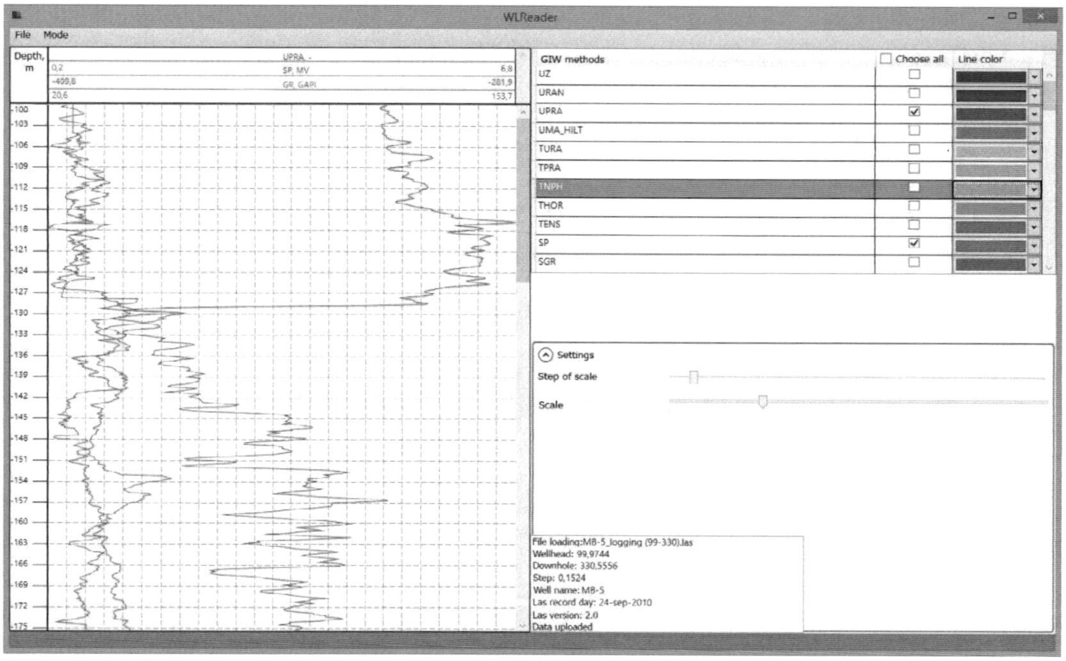

Figure 4 – Main program window

The program contains a function for combining the data of well logging interpretation and well logging (Fig. 5).

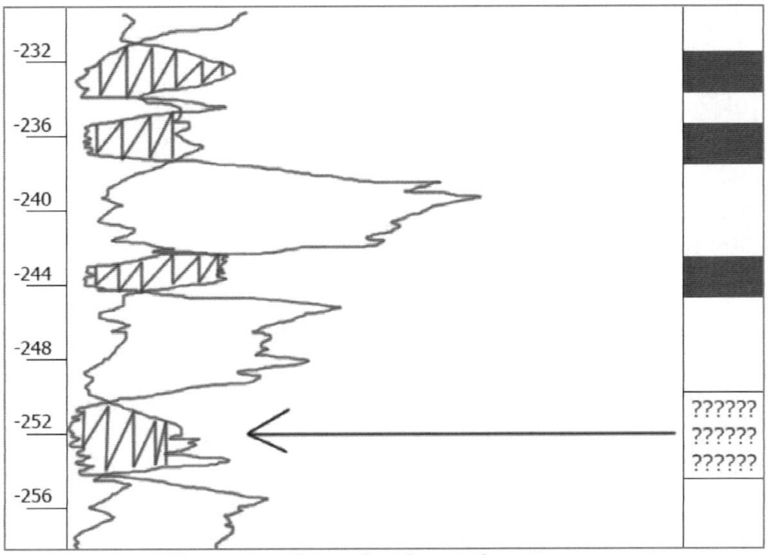

Figure 5 – Example of use of patterns

In Fig. 5, in the displayed image according to the results of combining the data, the brown color indicates the depths of oil occurrence. They all have intersecting plot lines resulting from two techniques. Thus, it can be assumed that the area with question marks can contain oil deposits.

3. Conclusion

As a result of this work, software for well logging data visualization and display of well log interpretation results has been developed. The developed software is able to display well logs and can be installed on the devices, working not only under Windows but Mac OS and Linux as well. It is also possible to run it on mobile devices under the Android operating system.

All functional and non-functional requirements formed at the design stage have been met.

The software under study is promising in terms of further development, which is achieved due to the realized possibility of new plugins development and connection to the program.

References

[1] Savinykh Y.A. Acoustic Method of Flow Rate Measurement in Multi-Layered Production / Y.A. Savinykh, H.N. Musipov // Avtomatizatsiya, Telemekhanizatsiya i Svyaz v Neftyanoy Promyshlennosti 2008. № 9. pp. 8-12.

[2] Musipov, H.N. Noiseproof Acoustic Channel in Well Drilling/ H.N. Musipov, M.A. Alexandrov, L.N. Bakanovskaya, M.V. Alferova // Avtomatizatsiya, Telemekhanizatsiya i Svyaz v Neftyanoy Promyshlennosti 2014. № 4. pp. 3-7.

[3] Akhmadulin, R.K. Cross-platform Software for Well Logging Data Visualization / R.K. Akhmadulin, L.N. Bakanovskaya, A.I. Miraev // Avtomatizatsiya, Telemekhanizatsiya i Svyaz v Neftyanoy Promyshlennosti 2017. № 1. pp. 44-48.

[4] Turenko, S.K. Adaptive Approach to Well Logging Data Processing in Building Seismic-Geological Models of Oil-Gas Facilities. / S.K. Turenko, E.A. Cherepanov // Neftyanoe Khozyaistvo 2016. № 5. pp. 34-37.

[5] Akhmadulin, R.K. The Development of the Software to Optimize Geophysical Field Oil and Gas Exploration / R.K. Akhmadulin, S.K. Turenko // IOP Conference Series: materials science and engineering. 2016. №1. Vol. 154. P. 012035. DOI: 10.1088/1757-899X/154/1/012035

[6] Dmitriev, A.N. Forward and Inverse Self-Potential Modeling: a New Approach / A.N. Dmitriev // Russian Geology and Geophysics. 2012. Vol. 53. № 6. pp. 611-622.
[7] Akhmadulin, R.K. Object-Oriented Programming When Developing Software in Geology and Geophysics / R.K. Akhmadulin, L.N. Bakanovskaya // IOP Conference Series: Earth and Environmental Science. 2017. №1. Vol. 50. P. 012049. DOI: 10.1088/1755-1315/50/1/012049
[8] Ferrari, C. LibHalfSpace: A C++ Object-Oriented Library to Study Deformation and Stress in Elastic Half-Spaces / C. Ferrari, M. Bonafede, M.E. Belardinelli // Computers and Geosciences. 2016. Vol. 96. P. 136-146. DOI: 10.1016/j.cageo.2016.08.011
[9] Zabolotin, A.E. Simulation of Triggering and Geodeformation Processes Induced by Fluid Injection Into the Geomedium / A.E. Zabolotin, A.V. Loskutov, A.V. Konovalov, S.B. Turuntaev // Russian Journal of Pacific Geology. 2016. Vol. 10, 6. P. 417-426. DOI: 10.1134/S1819714016060087
[10] Akhmadulin, R.K. An Object-Oriented Model of Case-Based Reasoning System Using Situations Tree / R.K. Akhmadulin, I.N. Gluhih, I.Y. Karyakin // Application of information and communication technologies, AICT2016. Conference proceedings: 12-14 October, 2016, Baku, Azerbaijan. IEEE, 2016. P. 124-127.
[11] Bakanovskaya, L.N. Improvement of Information Processing Quality in Preproduction Engineering Efficiency Assessment / L.N. Bakanovskaya, I.A. Chekardovskaya // IOP Conference Series: materials science and engineering. 2016. 1. 154. C. 012018. DOI: 10.1088/1757-899X/154/1/012018
[12] Glukhih, I.N. Digital Prototypes in Corporate Knowledge Bases of Well Interventions/ I.N. Glukhih, V.N. Piankov, A.R. Zabolotnev // Neftyanoe Khozyaistvo 2002. № 6. Pp. 45.
[13] Chekardovskaya, I.A. Modern Prototyping as a Basis for Industrial Design / I.A. Chekardovskaya, L.N. Bakanovskaya // IOP Conference Series: materials science and engineering. 2016. 1. 154. C. 012024. DOI: 10.1088/1757-899X/154/1/012024
[14] Bakanovskaya, L.N. Application of Artificial Neural Networks In the Heart Electrical Axis Position Conclusion Modeling / L.N. Bakanovskaya // IOP Conference Series: materials science and engineering. 2016. 1. 142. P. 012100. DOI: 10.1088/1757-899X/142/1/012100

IOP Conf. Series: Materials Science and Engineering **221** (2017) 012006 doi:10.1088/1757-899X/221/1/012006

Problems of Automation and Management Principles Information Flow in Manufacturing

E N Grigoryuk [a] **and V V Bulkin** [b]

Murom Institute of Vladimir state University, Orlovskaya, str., 23, Murom, 602264, Russia

E-mail: [a] kat-grigoryuk@yandex.ru; [b] vvbulkin@mail.ru

Abstract. Automated control systems of technological processes are complex systems that are characterized by the presence of elements of the overall focus, the systemic nature of the implemented algorithms for the exchange and processing of information, as well as a large number of functional subsystems. The article gives examples of automatic control systems and automated control systems of technological processes held parallel between them by identifying strengths and weaknesses. Other proposed non-standard control system of technological process.

1. Introduction

For several decades much attention is paid to study the creation and implementation of automatic control systems (ACS) and automated control systems of technological processes (ACSTP), which is one of the ways to improve production management and increase its effectiveness. High level of automation of technological objects allows increasing the quality and quantity ratio. ACSTP is intended primarily for development and implementation of control actions on the technological object of management. The goal is comparing the performance of ACS and ACSTP, analysis of management principles, as well as structuring and formalization of these principles. The non-standard control system of TP is offered.

2. The standard system of automatic control

A major part of control theory is the theory of automatic regulation. There is a standard system of automatic regulation (SAR) consisting of the control object, sensor and controller. Standard block diagram SAR shown in figure 1. For SAR affected by external factors (the disturbance $f(t)$) tending to bring it out of balance. Perturbations are: switching, harmonic, stepped, stochastic, white noise [1, 2].

The goal of SAR is to maintain a constant parameter or change it for a given algorithm, with adjustable size is slightly different from the original values.

Using some well-known approaches to the analysis of systems or devices auto-compensation of the interference influence [3], should pay attention to two principles of control – control perturbation (CP) and the control deviation (CD) [4].

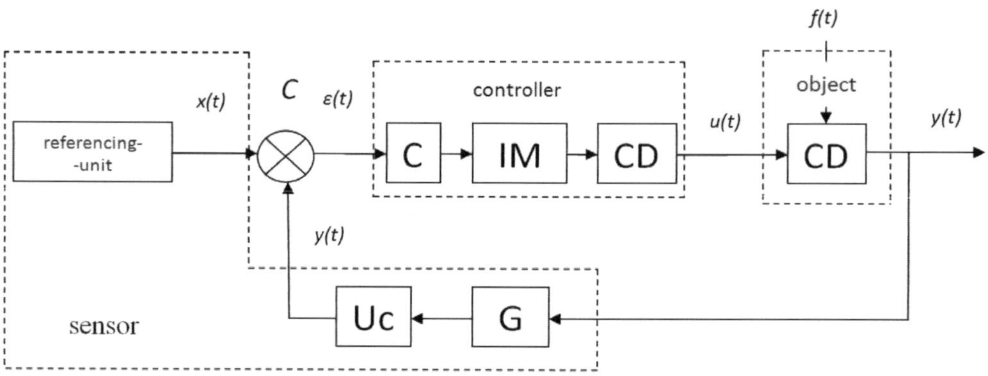

Figure 1. Standard block diagram SAR

The essence of the principle based regulation by the disturbance or compensation) based on the allocation of the total number of perturbations *f(t)*, the chief responds to SAR. Compensation to the control parameter *y(t)* of the fundamental disturbance, and the regulatory impact *u(t)* is formed in the SAR, based on the measurement results of the main disturbing factor acting on the object [5]. Block diagram SAR Poncele presented in figure 2.

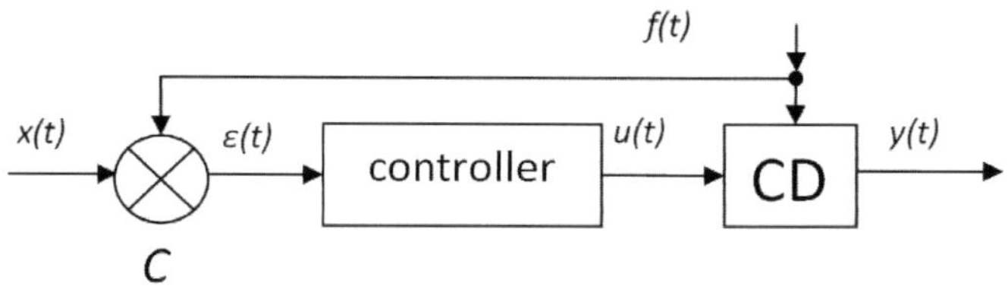

Figure 2. Block diagram SAR Poncele
$\varepsilon(t)=x(t)-y(t)$ – misalignment

Advantages:
- possible full invariance to certain disturbances;
- the stability of the system in the absence of feedback.

Disadvantages:
- a large number of perturbations requires a corresponding number of compensation channels;
- when changing the parameters of the controlled object causes errors in management;
- ability to apply only in clearly designated the characteristics of the object.

The essence of the principle of regulation according to the deviation, based on any deviation of the controlled parameter *y(t)* from its initial value *x(t)*, leads to the formation of the regulatory impact *u(t)* irrespective of the nature and number of disturbances. Block diagram SAR Polzunov–Uatt depicted in figure 3.

Consider the controlled parameter *y(t)* is the comparison with the set point *x(t)* defined on the misalignment $\varepsilon(t)=x(t)-y(t)$ and the controller produce control action *u(t)* to generate the control action requires the existence of an error) [5].

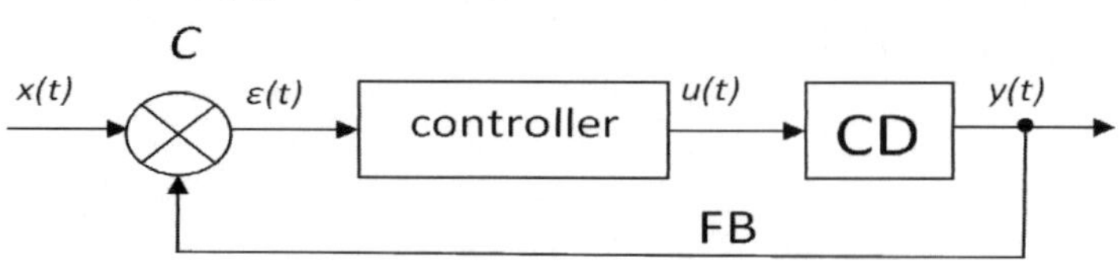

IOP Conf. Series: Materials Science and Engineering **221** (2017) 012006 doi:10.1088/1757-899X/221/1/012006

Figure 3. Block diagram SAR Polzunov–Uatt
FB – feedback

Advantages:
- negative feedback reduces the error regardless of factors it caused;
- less sensitive to variation of parameters of elements of the system, compared to control external influences (disturbances).

Disadvantages:
- there is a problem of sustainability;
- in simple single-loop systems it is impossible to achieve absolute invariance.

3. Combined control

The combined principle (closed–open loop principle) is a symbiosis of the principle Poncele and the principle of Polzunov–Uatt, combining their advantages and disadvantages. A block diagram of a combined SAR is shown in figure 4.

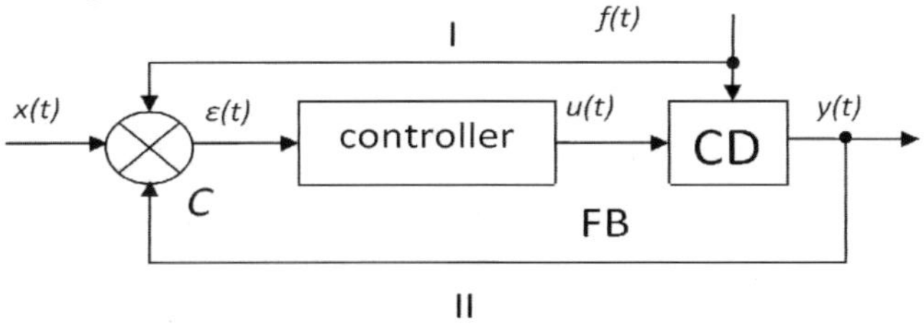

Figure 4. Block diagram of a combined SAR

Advantages:
- the presence of negative feedback makes the system less sensitive to changes in the parameters;
- add any channels (sensitive to job or resentment) does not affect the stability of the feedback loop.

Disadvantages:
- hindered the practical implementation of channels sensitive to job or indignation because they contain differentiating units;
- the crossing does not allow for all objects.

The impact load *f(t)* is provided by the instantaneous formation of the regulatory impact *u(t)*, given change of load (circuit I) and the difference (circuit II), for correction of errors arising due to inaccurate throttling under load [6].

4. Comparison of industrial automation systems

ACSTP are qualitatively different from SAR.

SAR is intended to stabilize the modes, processes and units. SAR, as a rule, is a closed management system, operating without personnel. The main purpose of the job optimally, ensuring the stabilization of the technological parameter or the desired physical quantity, wherein the reference value is always known and can be variable or constant, but of a known law [7, 8].

In contrast to SAR, the structure of the process control system involves the participation of the operator in making management decisions, including the structure of the outline of the formation of operator control actions because a main goal of process control system is the implementation of the optimal operation mode [9]. Industrial automation system is depicted in figure 5.

Criteria of optimality of technological regimes are the technical and economic performance (efficiency, specific consumption of raw materials, energy, fuel, etc.), obtained in the process of computational procedures and typically cannot be changed.

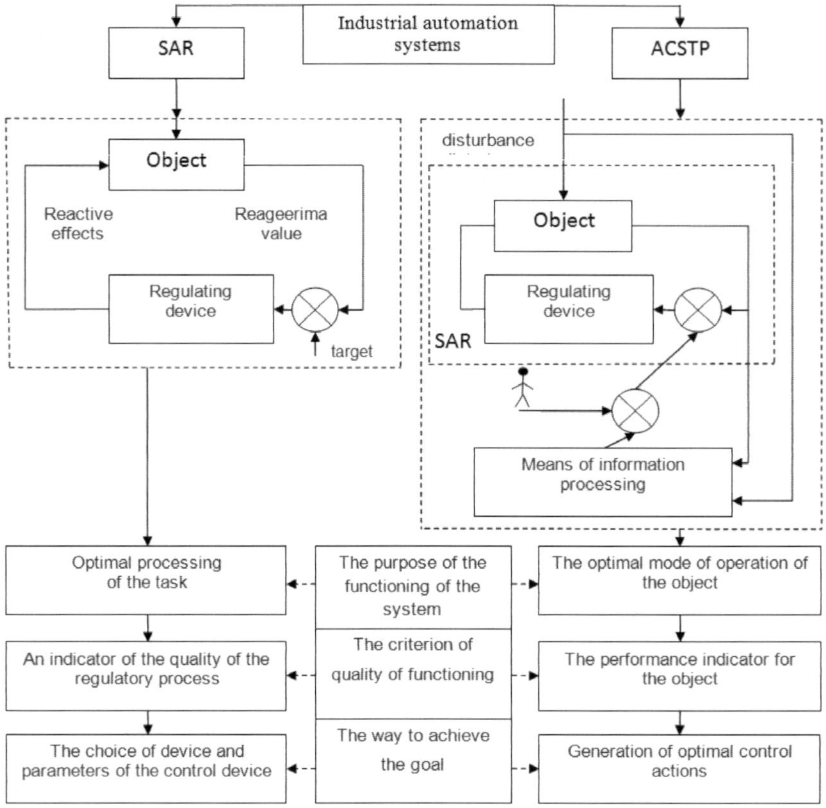

Figure 5. Industrial automation systems

5. Control system of complex technological process

The proposed control system (figure 6) is based on the structural study of the technological process [10]. At the entrance to the system of material and energy resources provided auxiliary production. Based on the theory of auto-compensation of interfering influences, the description of requirements represented as:

$$P = U_0(t) \cdot e^{j[\omega \cdot t + \varphi_0(t)]}, \tag{1}$$

where $U_0(t)$ – is a quantitative characteristic of the reference stream; ω – qualitative characteristic; $\varphi_0(t)$ – is the phase (temporary deviation) of the reference flux; t – is time; j – is an imaginary unit.

Components of physical operations Q ($Q = \{A,E,C\}$, where A and C – respectively the inlet and outlet streams; E – conversion operation A in C) expressed in terms of the ratio.

$$A = U_1(t) \cdot e^{j[\omega \cdot t + \varphi_1(t)]} \tag{2}$$

and

$$C = U_2(t) \cdot e^{j[\omega \cdot t + \varphi_2(t)]}, \tag{3}$$

where $U_1(t)$ and $U_2(t)$ – qualitative characteristics, respectively, the input and output streams; $\varphi_1(t)$ and $\varphi_2(t)$ – phase (temporary deviations) of the input and output streams.

The main (Central) process is a production process (conversion process), E is expressed via the ratio of the quantitative-qualitative conversion $K(p,u,\varepsilon)$

$$E = K(p, u, \varepsilon), \tag{4}$$

where ε – external destabilizing factors (the factors of deceleration, changes of plans under the pressure of external circumstances, etc.) $p=d/dt$ – Laplace operator, which allows to take into account the inertia of the system; u – control action.

Then (3), subject to (4) can be represented in the form

$$C = U_2(t)\cdot e^{j[\omega\cdot t+\varphi_2(t)]} = E\cdot U_1(t)\cdot e^{j[\omega\cdot t+\varphi_1(t)]}. \tag{5}$$

Of the comparison in the control system represents the reference comparison with the actual course of the process. The output is the product to meet quality standards.

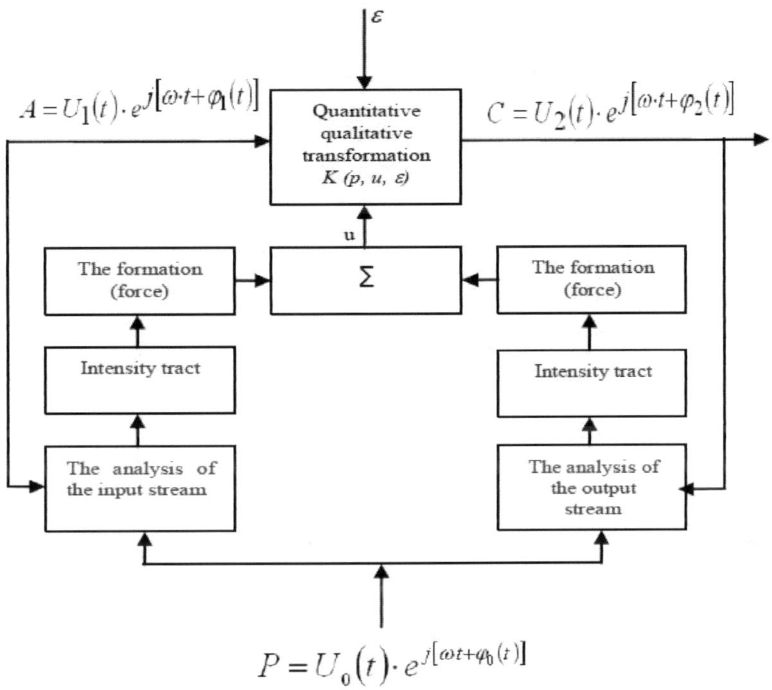

Figure 6. Control system of technological process

6. Summary

High level of automation of technological objects allows increasing the quality and quantity ratio. The principles of process management are based on understanding and analysis of input and output information flows and reflect a common management approach. The introduction presents the control system of complex technological process will allow to increase productivity, improve product quality, to create conditions for saving of material and labor resources, and optimization of production costs. Based on this analysis, it is possible to build more complex multi-level system, exciting in his description of all the strata process.

References

[1] Volchkevich L. Automation of production processes (Russia: Publ: mechanical engineering) p 380
[2] Besekersky V. 2007. The theory of automatic control systems (Russia, Moscow: Profession) p 752

[3] Kurilov I., Romashov V., and Popov, P. 1984 The system of compensation in phase and amplitude in the measuring devices. Automation geomagnetic studies (Russia, Moscow: Science) pp 145-155.
[4] Kurilov I., Grigoryuk E., Kalinichenko M., Kirillov I., Lashin A., and Bulkin. 2013 Principles for the management of information flows in industrial processes *Methods and devices of transmitting and processing information* (Russia), 15 pp 60-65
[5] Miroshnik I. 2005 Automatic control Theory. The linear system (Russia: St. Petersburg Publ: Peter) pp 336
[6] Bulkin V. and Grigoryuk E. 2014 The control System of complex technological process of the industrial enterprise *Collection of scientific papers of the XVIII International scientific and practical conference "System analysis in design and management* (Russia: St. Petersburg, Russia) pp 147-149
[7] Edited Egupov N. Methods of classical and modern automatic control theory. 2000 vol. 2: controller Synthesis and optimization theory, automatic control systems (Russia, Moscow Publ: MGTU im. N. Uh. Bauman) p 736
[8] Konovalov, B. and Lebedev, Y. 2010 Theory of automatic control (Russia, Tomsk TUSUR) p 162
[9] Potapova T. 1999 Structural analysis of a control system of a continuous closed production equipment and control systems (Russia), 12 pp 16-24
[10] Grigoryuk E. and Bulkin V. Principles of management manufacturing process of the primer-igniter on the basis of simulation *Proceedings of 2015 International Conference on Mechanical Engineering, Automation and Control Systems*, MEACS 2015. DOI: 10.1109/MEACS.2015.7414859.

IOP Conf. Series: Materials Science and Engineering **221** (2017) 012007 doi:10.1088/1757-899X/221/1/012007

Mathematical Description and Modeling of the Vibration Isolation Device with Neodymium Compensator Stiffness

E G Gurova

Mechatronics and automation department, Novosibirsk State Technical University, Av. K. Marx, 20, Novosibirsk, 630073, Russia

E-mail: lena319@mail.ru

Abstract In this article a mathematical description of the block diagram of the vibration isolation device with stiffness compensator is given. The vibration isolation system simulation performed with different functional elements, during which operation waveforms obtained with neodymium device compensator of the stiffness. Research & Development is under the scholarship of the President of Russian Federation, order No 184 from 10th of March 2015.

Introduction

To obtain the properties of the vibration isolation devise with neodymium compensator of the stiffness expediently investigate its mathematical model on theory stage. The equation of the movement of the mass, which is connected with basis of the vibration isolation device with compensator of the stiffness (without adjusting system), has been obtained from mathematical model of the device. In [1, 2, 3] the equation of the vibration isolation device with compensator of the stiffness (linear and nonlinear) has been obtained also. High requirements for precision and reliability of the operation of the mechanical transport system are fundamental issue of the development. The stability of the device and separated parts operation is impacted by dynamic loads during of the exploitations. The sources of these disturbances can be electromechanical devices [4, 5]: drives, actuating devices (fans, propellers) and etc.

In case of the using of the vibration isolation devices of the nonlinear compensators, the investigation must be performed with the nonlinear systems analysis methods.

Calculation and simulations

Differential equation is described the behavior of the mass on elastic suspension, with spring and nonlinear stiffness compensator has a view:

$$mx'' + cx - F_1(x) + F_{sp} = F(t), \tag{1}$$

where m is mass of the object;
x is movement of the object;
c is rigidity of the elastic element;
$Fc\ (x)$ is function, which is described the characteristic of the compensator;
$k_{sp}\ (x)$- is coefficient of the control system, depended from movement of the oscillations center of the anchor;

F(t) is disturbing force.

With basis of the equilibrium equation (1) we can make cinematic scheme of the device with compensator of the stiffness and control system [6, 7, 9]. The view of the scheme is shown on figure 1.

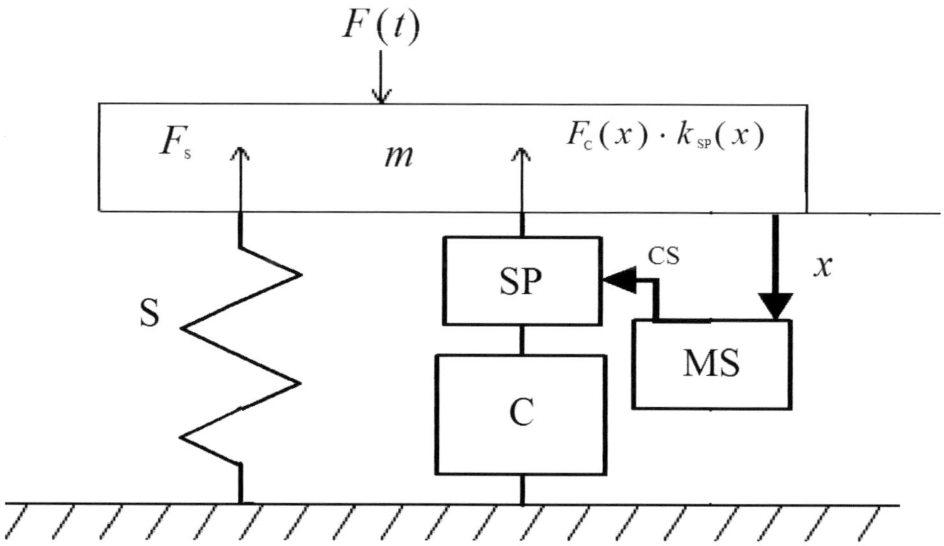

Figure 1. Cinematic scheme of the vibration isolation suspension with the stiffness compensator and control system

The equation of the traction characteristic of the supermagnet compensator [10, 13] can be represented as:

$$F(x) = F_1 + F_2 = \frac{\pi\mu_0}{4}\left(\frac{2B_0}{\mu_0}\right)^2 R^4\left(\frac{1}{(x)^2} + \frac{1}{(x+2h)^2} - \frac{2}{(x+h)^2}\right) -$$

$$-\frac{\pi\mu_0}{4}\left(\frac{2B_0}{\mu_0}\right)^2 R^4\left(\frac{1}{(b-x)^2} + \frac{1}{(b-x+2h)^2} - \frac{2}{(b-x+h)^2}\right), \quad (2)$$

where $\mu_0 = 4\pi \cdot 10^{-7}$ is permeability, *H/m*;
δ is max distance between magnets, *m*;
B_0 is residual magnetic induction of the magnet, *T*;
R is radii of the magnet, *m*;
h is height of the magnet, *m*;
x is shift from the center, *m*.

For compilation and theoretical investigation of the models of the vibration isolation devices is much convenient to use the method, which is used in theory of the automatic control (representation of systems as structural schemes and its modelling [7, 9]).

When compiling a design scheme in the program Simulink adhere to some rules:

- each unit has ony input value and one output value (in system of the automatic control the value usually are named variables);
- all units are single directed, i.e. input variable impacts on output variable, but output variable on input variable does not;

IOP Conf. Series: Materials Science and Engineering 221 (2017) 012007 doi:10.1088/1757-899X/221/1/012007

- the units are described by transfers function instead of differential equations. The transfer function is the ratio of the Laplace image of the output signal and input signal with zero initial conditions.

On he figures 2 and 3 the mathematical model of the oscillation system without stiffness compensator and oscillogram of resulting shifts of protected object, consequently, are shown.

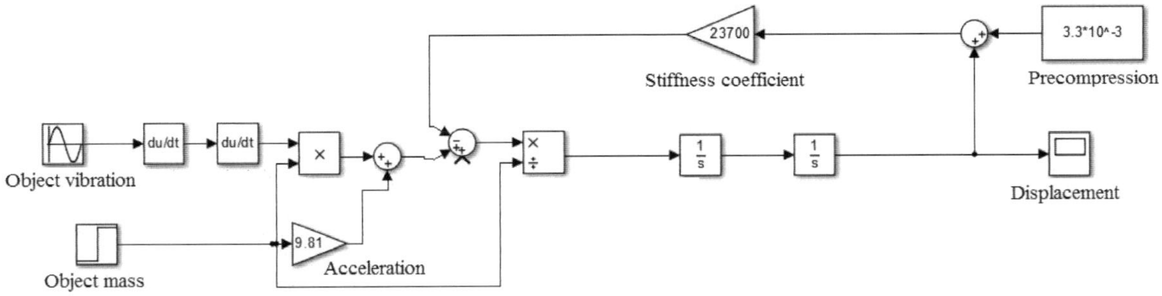

Figure 2. The mathematical model of the oscillation system with rigid element and without stiffness compensator

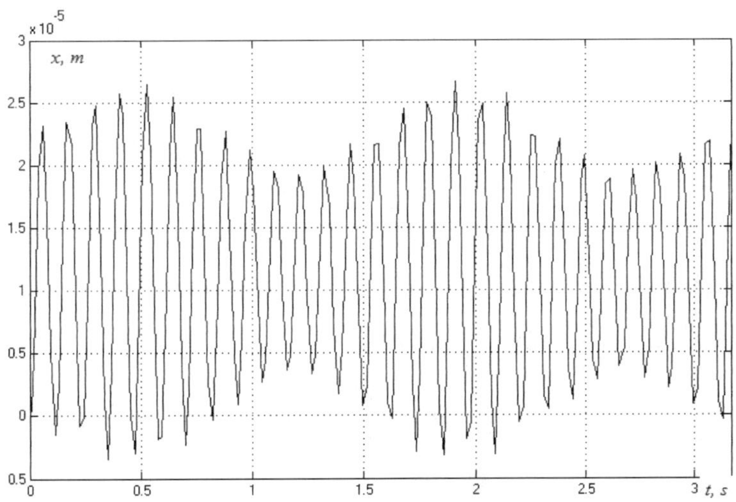

Figure 3. Oscillogram of the oscillation system operation without stiffness compensator

The simulation has been shown that system has non sustainability oscillations. Inside of the system there are established oscillations of the different frequencies and amplitudes, and max amplitude is achieved 3 *sm*. These oscillations are an unsatisfactory indicator of the vibration isolation system.

To realize full model of the vibration isolation device the connection of the stiffness compensator is necessary [7, 11]. The view of the vibration isolation suspension model with neodymium compensator of the stiffness is shown on figure 4.

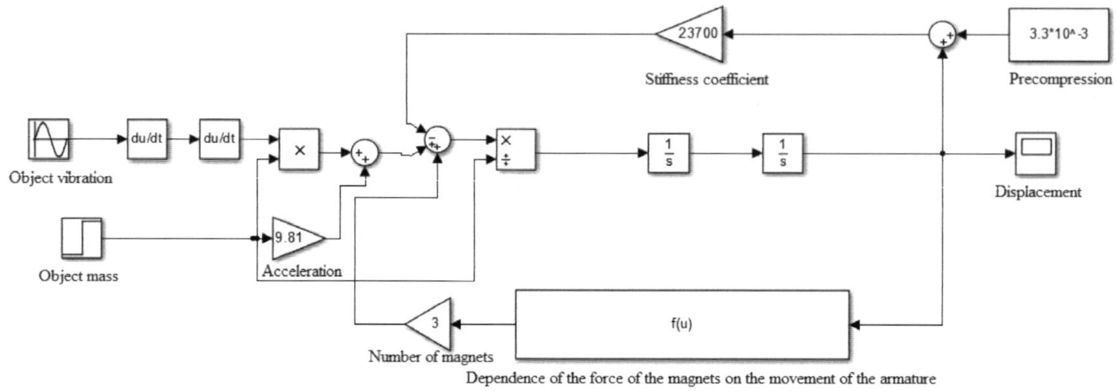

Figure 4. Mathematical model of the vibration isolation device with compensator of the stiffness

This structural scheme allows simulate the behavior of the object under any disturbances *F(t)*, and in the model it is possible to change the stiffness of the basic elastic element *c* and the compensator c_k and the object mass *m*. However in this model does not take to account the presence of the control system, which is tracking the changes of the static part of the force *F(t)*, i.e., the model allows investigation of the behavior of the system only in stationary regime.

The initial data of the model of the vibration isolation device:

- mass of the object $m = 8\ kg$;
- stiffness of the spring $c = 2370\ N/m$.

For represented model the input signal is sine wave with parameters:

- amplitude - *1.5 mm*;
- frequency $2\pi \cdot 8$ *rad/sec*;
- another parameters are *zero*.

As the load the step-signal is used. The load is not connected to vibration isolation system without correcting unit.

The results of the simulation are shown on figure 5.

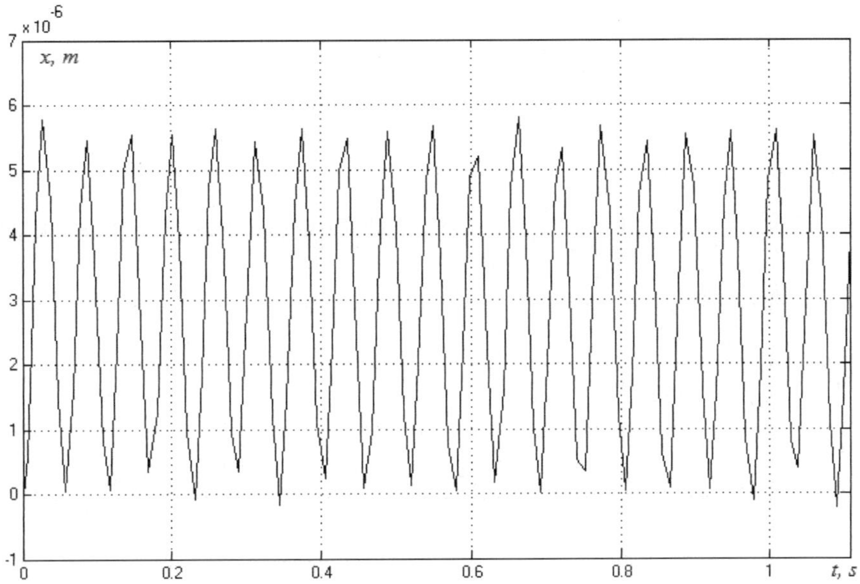

Figure 5. Oscillogram of the vibration isolation device with neodymium stiffness compensator operation

From results of the simulation we can see that there are conservative oscillations in the system with a sufficiently small amplitude of the order of a thousandth of a millimeter. However, the oscillations presence in the positive displacement region, which indicates that the compensators force acts with a larger value than the elastic force, which caused displacement of the oscillation axis.

To obtain sustainable oscillations of the vibration isolation device and possibility of the correction of its properties [8, 12], we need to include inside the model the control system (figure 6). In this case oscillogram of the resulting shift became a view, is shown of figure 7.

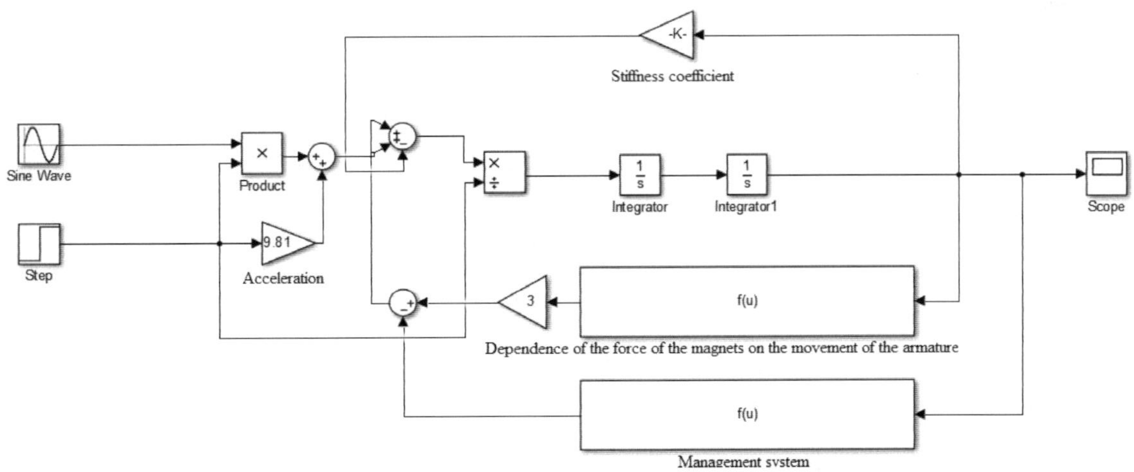

Figure 6. Mathematical model of the vibration isolation device with compensator of the stiffness and with adjusting system

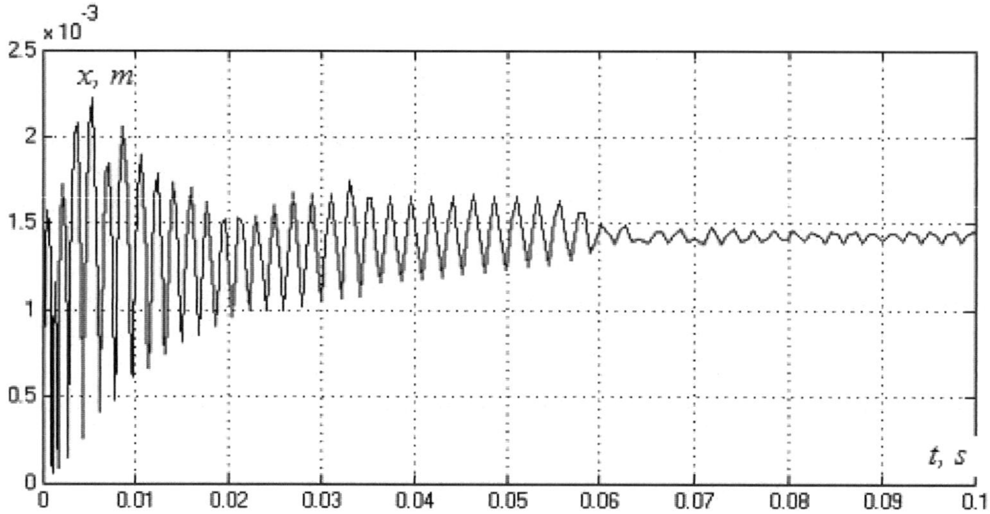

Figure 7. Oscillogram of the vibration isolation device operation with neodymium compensator of the stiffness and control system

The results of the simulation allow us to make the conclusion, that vibration isolation system with neodymium compensator of the stiffness is sustainable with inclusion of the control system.

Results

As result of scientific research, which were led within the framework of development of the vibration isolation devices with the stiffness compensators, a mathematical model is presented, on the basis of which simulation of vibration isolation systems was performed. The basis of the research was software Matlab 6.5 Simulink. The simulation has shown that the system with the stiffness compensator based on neodymium magnets is very effective for decreasing of the vibration oscillation under different load when adjusting system is connected.

Conclusion

The research «Vibration isolation devices with the stiffness compensators based on electromagnets anв neodymium magnets» was performed under the scholarship of the President of Russian Federation for young scientists, order No 184 from 10th of March 2015.

References

[1] Alabuzhev P M, Zuev A K, Negodaev M N, Yarunov A M, Kargin V A, Kopeikin G F and Khan V B 1965 Use of high-speed cinematography to study the action of an electromechanical detent hammer *Journal of mining science* **2 № 1** pp 139-141, doi: 10.1007/BF02501941

[2] Alabuzhev P M, Gritchin A A, Stepanov P T, Khon V F 1977 Some results of an investigation of a vibration protection system with stiffness correction Soviet mining science USSR 13 № 3 pp 338-341, doi: 10.1007/BF02511213, (WOS:A1977EY54900024)

[3] Bonisoli E, Vigliani A 2007 Identification techniques applied to a passive elasto-magnetic suspension *Mechanical systems and signal processing* **21 № 3** pp 1479-1488, doi: 10.1016/j.ymssp.2006.05.009 (WOS:000243590300019)

[4] Gurova E G and Gurov M G 2014 Vibro isolator with neodymium magnets compensator of the stiffness *Applied Mechanics and Materials* **682** pp 118-121, doi: 10.4028/www.scientific.net/AMM.682.118)

[5] Gurova E G, Gurov M G , Makarov S V and Sergeev A A Features of the power characteristics

of the vibration isolators *Advanced Materials Research* 1040 pp 678-681, doi: 10.4028/www.scientific.net/AMR.1040.678 (000348897100127)
[6] Gurova E G 2015 Improving of the operation efficiency of the vehicle due to using of the neodymium magnets inside the vibration isolation devices *Conf. Series-Materials Science and Engineering* **91** 012091, doi: 10.1088/1757-899X/91/1/012091 (WOS:000361836000091)
[7] Gurova E G, Gurov M G and Panchenko Y V 2016 Simulation of the Magnetic Characteristics and Properties of the Neodymium Compensator of the Stiffness *IOP Conference Series: Materials Science and Engineering* **142, № 1** 123667, doi: 10.1088/1757-899X/142/1/012131
[8] Ishida S, Suzuki K and Shimosaka H 2016 Design and experimental analysis of origami-inspired vibration isolations with quasi-zero-stiffness characteristic *Proceedings of the asme international design engineering technical conferences and computers and information in engineering conference* **5B** V05BT07A015 (WOS:000393365000015)
[9] Prisekin V L, Pustovoi N V and Rastorguev G I 2014 Algorithms for controlling fatigue tests of airplanes *Journal of Applied Mechanics and Technical Physics* **55 № 1** pp 164-171, doi: 10.1134/s0021894414010192 (WOS:000331713200019)
[10] Rivin E I 1995 Vibration isolation of precision equipment *Precision engineering – journal of the american society for precision engineering* **17 № 1** pp 41-56, doi: 10.1016/0141-6359(94)00006-L (WOS:A1995QJ99100006)
[11] Spiewak S, Karpachevskyy T and Zaiss C 2014 High performance isolation and excitation of vibration for enhanced identification of inertial sensors *Proceedings of the asme international mechanical engineering congress and exposition* **4B** V04BT04A061 (WOS:000359682700061)
[12] Wang J, Zhao S G, Wu D F and Jing X J 2016 The interior working mechanism and temperature characteristics of a fluid based microvibrationisolator *Journal of sound and vibration* **360** pp 1-16, doi: 10.1016/j.jsv.2015.09.008 (WOS: 000363187400001)
[13] Xu D L, Yu Q P, Zhou J X and Bishop S R 2013 Theoretical and experimental analyses of a nonlinear magnetic vibration isolator with quasi-zero-stiffness characteristic *Journal of sound and vibration* **332 № 14** pp 3377-3389, doi: 10.1016/j.jsv.2013.01.034 (WOS:000318579700002)

IOP Conf. Series: Materials Science and Engineering **221** (2017) 012008 doi:10.1088/1757-899X/221/1/012008

Realization of The Adjusting System in Stiffness Compensator Based on Permanent Neodymium Magnets

E G Gurova

Mechatronics and automation department, Novosibirsk State Technical University, Av. K. Marx, 20, Novosibirsk, 630073, Russia

E-mail: lena319@mail.ru

Abstract In this paper one of the cases of adjusting system for the stiffness compensator based on permanent magnets with changing of the axial distance between magnets has been described. The dependency of the traction force from value of the rotation angle of the anchor relatively basis in stiffness compensator has been described. Research & Development is under the scholarship of the President of Russian Federation, order No 184 from 10th of March 2015.

Introduction

In recent time, the level of the vibration from power installations has been increased up to high values. The negative impact of the vibration from vibrating installations influences on design, the equipment, electronics and people, which are using of these installations.

As practice shows, the most promising means of fighting vibration is the inclusion of the vibration isolators, that is, suspension of the equipment with source of the vibration [1, 2, 4, 11, 17- 20, 21]. This method allows us significantly reducing of the vibration level. However, vibration isolators are less effective against low-frequency oscillations, which are caused by imperfection of their design.

The use in the vibration isolator design of the vibration isolator will significantly increase the efficiency of the suspension device [2, 3, 12-16, 22]. The vibration isolator has a traction characteristic having a negative slope, but in absolute value equal to the characteristic of the elastic element, which is included in the vibration isolation suspension.

In [6, 8, 9, 10], the authors proposed the installation of a vibration isolator which is equipped with the stiffness compensator based on electromagnets. In [6, 7], the issue of reducing the size of the device, the authors has been proposed the replacement of the electromagnetic coils with permanent neodymium magnets. However, a necessary condition of the absolute vibration isolation is a adjusting system, which allows the adjustment of the stiffness compensator to the external impact to provide full vibration isolation under different initial conditions.

It should be noted that the adjustment system of the stiffness compensator which has electromagnets as the basis, was easy to realization, as it is possible to regulate the power supply of electromagnets and to change the traction caused by their interaction. As for the neodymium compensator, its adjustment system remains far from obvious. It is necessary to choose the most suitable way to adjust the stiffness compensator.

Calculation and simulations

In this paper the authors have been proposed to use the adjustment system of the stiffness compensator based on neodymium magnets with rotation anchor with magnets. The core of this approach is the transmission of the position sensor signal of the protected and oscillated objects to device that drives the stock with anchor. The anchor with installed magnet is rotating and changes the distance between the centers of the magnets because of that the force of the interaction between the magnets of the anchor and basis of the stiffness compensator is changed.

In this paper the characteristic of the traction force is obtaining in the stiffness compensator with in dependency with the anchor rotation angle has been considered. From the equation presented in [10], the force of the interaction between the magnets from its shift is possible to define from next dependency:

$$F(r) = \frac{\pi}{2\mu_0} {B_0}^2 R^4 \sum_{i,j=0}^{1} \frac{(-1)^{i+j}}{(x+h(i+j))^2} \left(1 - \frac{3}{2} \frac{r^2}{(x+h(i+j))^2}\right) \quad (1)$$

The (1) shows the dependency of the traction force form axial distance between two magnets.

Three counter directed magnets are supposed to use in design of the stiffness compensator. The magnets are placed to attract to each other (see figure 1).

Dependency of the force arises inside of the stiffness compensator with shift of the magnets:

$$F(r) = \frac{\pi}{2\mu_0} {B_0}^2 R^4 \sum_{i,j=0}^{1} \frac{(-1)^{i+j}}{(x+h(i+j))^2} \left(1 - \frac{3}{2} \frac{r^2}{(x+h(i+j))^2}\right) - \frac{(-1)^{i+j}}{(b-x+h(i+j))^2} \left(1 - \frac{3}{2} \frac{r^2}{(b-x+h(i+j))^2}\right), \quad (2)$$

where r is distance between center of the magnets;

B_0 is residual magnetic induction of the magnet;
R is radii of the magnet;
h is height of the neodymium magnet;
x is magnitude of the gap between of the anchor and the basis.

The compact dependency is obtained by this way is sum of the double series [10, 12], that is the sums of the set of the numbers to the elements of which the positive integers *i* and *j* are assigned.

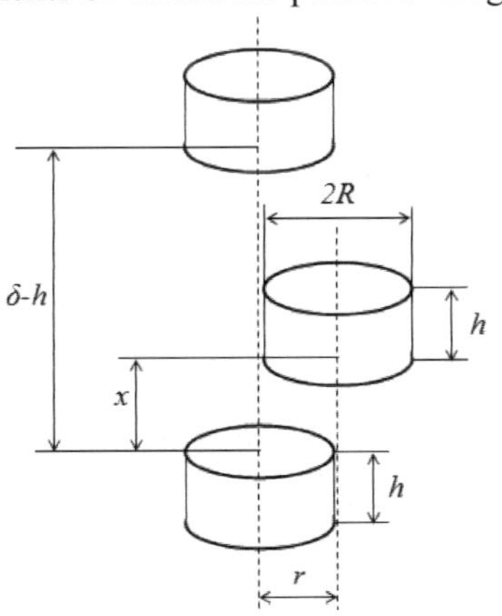

Figure 1. The sketch of the placement of the anchor and basis of the stiffness compensator

IOP Conf. Series: Materials Science and Engineering **221** (2017) 012008 doi:10.1088/1757-899X/221/1/012008

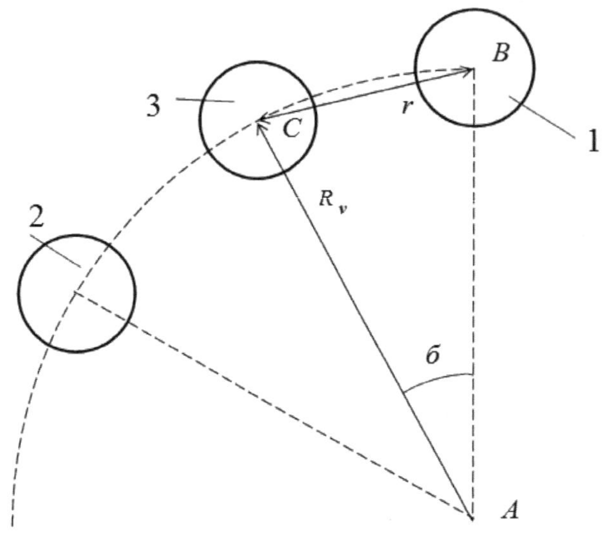

Figure 2. The sketch of the placement of the magnets from the top

It can be seen form figure 1 that the magnets 1 and 2, are placed in the base of the stiffness compensator are interacting with movable anchor of the device through the magnet 3. Thus, starting form the triangle *ABC* the dependency of the distance between the magnets from rotation angle of the device anchor can be establishing.

The distance between the magnets has dependency from rotation angle of the anchor accordingly the cosine theorem:

$$r(\alpha) = \sqrt{2R_y(1\text{-}\cos(\alpha))} \tag{3}$$

where R_y is the radii of the placement of the anchor magnets.

It is useful to note that all magnets of the device impact on the anchor magnets. However the design of the vibration isolator has been made to minimize impact from the neighbor magnets on each other, and because its impact can be neglected. As can be seen on figure 1 the strongest impact on anchor magnet with some rotation angle are made two neighbor magnets of the basis 1 and 2 in the moment of its placement. Thus, it is necessary to take into account this impact on the anchor.

Staring from the condition that magnets which placed in anchor and basis have the same characteristics, it is possible to write the force of its impact on the anchor magnet.

The dependency of the traction force between magnets from the anchor shift angle:

$$F(\alpha) = \frac{\pi}{2\mu_0} {B_0}^2 R^4 \sum_{i,j=0}^{1} \frac{(-1)^{i+j}}{\left(x + h(i+j)\right)^2} \left(2 - \frac{3R_y(2 - \cos(\alpha) - \cos(60 - \alpha))}{\left(x + h(i+j)\right)^2}\right) -$$
$$- \frac{(-1)^{i+j}}{\left(b - x + h(i+j)\right)^2} \left(2 - \frac{3R_y(2 - \cos(\alpha) - \cos(60 - \alpha))}{\left(b - x + h(i+j)\right)^2}\right). \tag{4}$$

Accordingly (4) the dependency of the traction force is acting on the anchor magnet from rotation angle has been drawn:

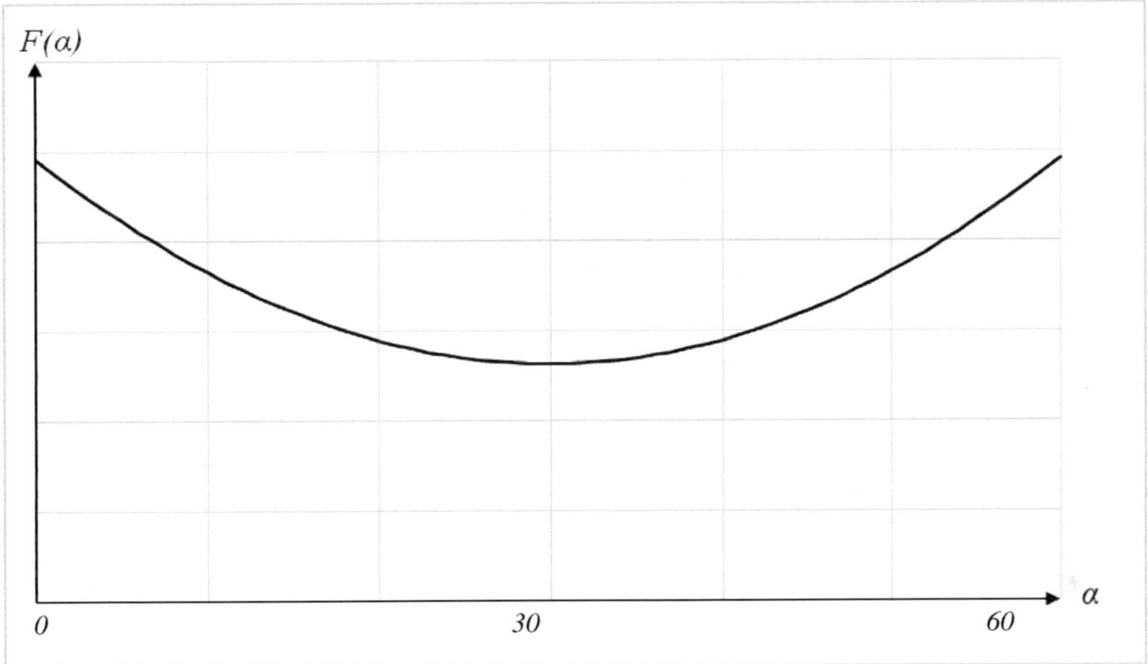

Figure 3. Dependency of the force from the rotation angle of the anchor

It can be seen from figure 3 the max value of the angle is angle between two magnets in the basis of the stiffness compensator (*60°*). Max values of the force of the interaction of the anchor and the basis of the stiffness compensator are angles (*0°* и *60°*) where the magnets of the anchor and the basis are placed on the same axis. Also from this plot can be seen that the minimum force achieves by the magnets when the magnets of the anchor are placed between magnets of the stiffness compensator basis. Thus, it can be concluded that minimum force between the magnets is almost half less than maximum achieved force.

Conclusion

From results of this dependency it was established that when the rotation angle changes it is possible to change the traction force and consequently the slope of the traction characteristic of the stiffness compensator. It is allowing the vibration isolation system to adjust to the level of the oscillations and thus allow the effectively isolate it. Further research the proposed adjustment system can be realized in design of the vibration isolator.

The research «Vibration isolation devices with the stiffness compensators based on electromagnets anв neodymium magnets» was performed under the scholarship of the President of Russian Federation for young scientists, order No 184 from 10th of March 2015.

References

[1] Alabuzhev P M, Zuev A K, Negodaev M N, Yarunov A M, Kargin V A, Kopeikin G F and Khan V B 1965 Use of high-speed cinematography to study the action of an electromechanical detent hammer *Journal of mining science* **2 #1** pp 139-141, doi: 10.1007/BF02501941

[2] Alabuzhev P M, Gritchin A A, Stepanov P T, Khon V F 1977 Some results of an investigation of a vibration protection system with stiffness correction Soviet mining science USSR 13 № 3 pp 338-341, doi: 10.1007/BF02511213, (WOS:A1977EY54900024)

[3] Bartel T, Gaisbauer S, Stohr P, Herold S and Melz T 2014 Development of functionally integrated mounts for three- and six-axial vibration isolation of sensitive equipment *Proceedings of international conference on noise and vibration engineering (ISMA2014) and international conference on uncertainty in structural dynamics (USD2014)* pp 119-130 (WOS:000352201000009)

[4] Bonisoli E, Vigliani A 2007 Identification techniques applied to a passive elasto-magnetic suspension *Mechanical systems and signal processing* **21 (3)** pp 1479-1488, doi: 10.1016/j.ymssp.2006.05.009 (WOS:000243590300019)

[5] Carrella A, Brennan M J and Waters T P 2007 Static analysis of a passive vibration isolator with quasi-zero-stiffness characteristic *Journal of sound and vibration* **301 # 3-5** pp 678-689, doi: 10.1016/j.jsv.2006.10.011 (WOS: 000244294400016)

[6] Gurova E G and Gurov M G 2014 Vibro isolator with neodymium magnets compensator of the stiffness *Applied Mechanics and Materials* **682** pp 118-121, doi: 10.4028/www.scientific.net/AMM.682.118

[7] Gurova E G, Gurov M G , Makarov S V and Sergeev A A Features of the power characteristics of the vibration isolators *Advanced Materials Research* 1040 pp 678-681, doi: 10.4028/www.scientific.net/AMR.1040.678 (WOS: 000348897100127)

[8] Gurova E G, Gross V Y, Kurbatov V S, Makarov S V, Sergeev A A and Shchurov N I 2013 Development of spatial vibration protection devices *World Applied Sciences Journal* **22** pp 44-48, doi: 10.5829/idosi.wasj.2013.22.tt.22138

[9] Gurova E G 2015 Improving of the operation efficiency of the vehicle due to using of the neodymium magnets inside the vibration isolation devices *Conf. Series-Materials Science and Engineering* **91** 012091, doi: 10.1088/1757-899X/91/1/012091 (WOS:000361836000091)

[10] Gurova E G, Gurov M G and Panchenko Y V 2016 Simulation of the Magnetic Characteristics and Properties of the Neodymium Compensator of the Stiffness *IOP Conference Series: Materials Science and Engineering* **142, #1** 123667, doi: 10.1088/1757-899X/142/1/012131

[11] Ishida S, Suzuki K and Shimosaka H Design and experimental analysis of origami-inspired vibration isolations with quasi-zero-stiffness characteristic *Proceedings of the asme international design engineering technical conferences and computers and information in engineering conference* **5B** V05BT07A015 (WOS:000393365000015)

[12] Koyacic I, Brennan M J, Waters T P 2008 A study of a nonlinear vibration isolator with a quasi-zero stiffness characteristic *Journal of sound and vibration* **315 #3** pp 700-711, doi: 10.1016/j.jsv.2007.12.019 (WOS:000257516900019)

[13] Lin S Y, Xu Jand Cao H 2016 Analysis on the ring-type piezoelectric ceramic transformer in radial vibration *IEEE Transactions on power electronics* **31 #7** pp 5079-5088, doi: 10.1109/TPEL.2015.2482990 (WOS:000370528800034)

[14] Lee J, Ghasemi A H, Okwudire, C E and Scruggs J 2017 A linear feedback control framework for optimally locating passive vibration isolators with known stiffness and damping parameters *Journal of vibration and acoustics-transactions of the asme* **139 #1** 011006, doi: 10.1115/1.4034771 (WOS: 000391721800006)

[15] Lee C M, Goverdovskiy V N and Temnikov A I 2007 Design of springs with «negative» stiffness to improve vehicle driver vibration isolation *Journal of sound and vibration* **302 № 4-5** pp 865-874, doi: 10.1016/j.jsv.2006.12.024 (WOS: 000245508700015)

[16] Prisekin V L, Pustovoi N V and Rastorguev G I 2014 Algorithms for controlling fatigue tests of airplanes *Journal of Applied Mechanics and Technical Physics* **55 #1** pp 164-171, doi: 10.1134/s0021894414010192 (WOS:000331713200019)

[17] Rivin E I 1995 Vibration isolation of precision equipment *Precision engineering – journal of the american society for precision engineering* **17 #1** pp 41-56, doi: 10.1016/0141-6359(94)00006-L (WOS:A1995QJ99100006)

[18] Spiewak S, Karpachevskyy T and Zaiss C 2014 High performance isolation and excitation of vibration for enhanced identification of inertial sensors *Proceedings of the asme international*

mechanical engineering congress and exposition **4B** V04BT04A061 (WOS:000359682700061)

[19] Subrahmanyan P K, Trumper D L 2000 Synthesis of passive vibration isolation mounts for machine tools - *A control systems paradigm Proceedings of the 2000 American control conference* **1-6**, pp 2886 – 2891(WOS:000166106200592)

[20] Xu Z D, Xu F X and Chen H 2015 Vibration suppression on a platform by using vibration isolation and mitigation devices *Nonlinear dynamics* **83** pp 1341-1353, doi: 10.1007/s11071-015-2407-4 (WOS:000368688700016)

[21] Xu D L, Yu Q P, Zhou J X and Bishop S R 2013 Theoretical and experimental analyses of a nonlinear magnetic vibration isolator with quasi-zero-stiffness characteristic *Journal of sound and vibration* **332** **#14** pp 3377-3389, doi: 10.1016/j.jsv.2013.01.034 (WOS:000318579700002)

[22] Wang J, Zhao S G, Wu D F and Jing X J 2016 The interior working mechanism and temperature characteristics of a fluid based microvibrationisolator *Journal of sound and vibration* **360** pp 1-16, doi: 10.1016/j.jsv.2015.09.008 (WOS: 000363187400001)

IOP Conf. Series: Materials Science and Engineering **221** (2017) 012009 doi:10.1088/1757-899X/221/1/012009

Modeling of Delignification Process of Activated Wood and Equipment for its Implementation

D B Prosvirnikov, R G Safin, D F Ziatdinova, N F Timerbaev, A R Sadrtdinov

Institute of petroleum and chemical engineering, Kazan national research technological University, K. Marx Street 68, Kazan, Republic of Tatarstan, Russian Federation

E-mail: prosvirnikov_dmi@mail.ru

Abstract. The article is the analyses of the physical picture of the process of steam explosion treatment of pre-impregnated lignocellulosic material, on the basis of which the mathematical modeling is represented in the form of differential equations with boundary conditions. The obtained mathematical description allows to determine the degree of influence of various factors on the kinetics of the process.

1. Introduction

Delignification of activated wood is a two-stage combined pulping. Preliminary steam hydrolysis allows to remove most of the hemicelluloses from the wood and weaken the bonds between cellulose and pentosans, and the steam explosion due to the sharp decompression is intended to loosen the structure of the cell walls and partially separate the fibers from each other. These two interrelated phenomena make it possible to increase the reactivity of the resulting cellulose, which is intended for further chemical processing. The material after steam explosion treatment is a combination of cellulose, various lignin structures (native, low molecular weight, condensed), and hemicellulose residues in the form of mono- and oligosaccharides [1-4], and the products of degradation during steam explosion treatment are mono- and oligosaccharides, low-molecular and condensed lignin. Part of the native (original, natural) lignin remains unchanged.

2. Physicochemical picture of processes and phenomena occurring during delignification of activated wood

Before delignification, the purpose of which is to produce cellulose for further chemical processing, the blasted material is subjected to washing to extract destructured hemicelluloses (in the form of sugars) and low molecular weight lignin fragments (hereinafter, by-products of degradation). Lignocellulosic pulp after steam explosion treatment contains from 10 to 40% of by-products from the raw material, depending on processing conditions. From the total amount of products, a small part (approximately 5-7% of the total mass) is concentrated in the capillaries of the cell walls. 15-20% of the products of the total mass are enclosed in the channels and cavities of the fibers. 75-80% of the total mass, i.e. the remaining degradation products, constitute a free solution of the products in the water surrounding the individual fibers or bundles of fibers. After washing, the high-moisture fiber with a specified content of lignin and cellulose is loaded into the cooker, where the previously prepared delignifying agent is supplied. The material after the steam explosion treatment has an even

IOP Conf. Series: Materials Science and Engineering **221** (2017) 012009 doi:10.1088/1757-899X/221/1/012009

greater porosity due to the removal of easily hydrolyzable polysaccharides from it, as well as the highly developed specific surface formed as a result of the deflation due to the release of pressure. These properties contribute to a multiple acceleration of the penetration of the reagent into the cell wall of the wood material. The high humidity of the blasted fiber also helps to accelerate the diffusion of active substances.

The process of delignification of activated wood is greatly simplified compared to delignification of wood chips. The general mechanism of the delignification of activated wood and the sequence of the individual stages from which the transformation of lignocellulosic material into cellulose are as follows:

1) penetration of delignifying agents into the cell wall of the activated material, i.e. diffusion of the reagents (fig. 1) [5];

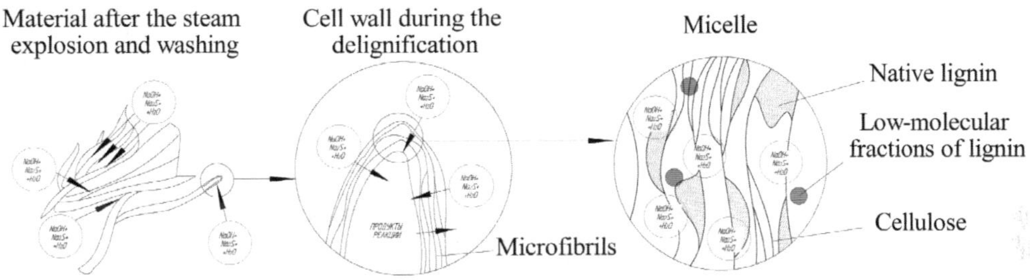

Figure 1. Mechanism of penetration of delignifying agents into the material obtained after steam explosion treatment of wood chips with subsequent washing

2) adsorption of active reagents in micelles on the high reaction surface of the activated material, accompanied by swelling of the cell walls [6];

3) stage of chemical reactions in the solid phase: interaction of the structures of phenylpropane units of lignin with the reagent, followed by a competing condensation reaction [7-9];

4) the final stage of chemical reactions: partial destruction of delignification products [10-11];

5) incomplete depolymerization and hydrolysis of hemicelluloses and cellulose [12-14];

6) further condensation or depolymerization of delignification products [15-16].

In the delignification process, a multicomponent mixture of vapors is formed, which consists predominantly of methyl alcohol, hydrogen sulfide, dimethyl sulphide and methyl mercaptan, turpentine, a small amount of acetone, methyl sulfur compounds, ammonia and ammonium sulphide. Therefore, taking into account the complexity of the phenomena of heat and mass transfer, in order to obtain a constructive mathematical description of a multistage process, we make the following assumptions:

1. We assume that the particle after the steam explosion treatment has the shape of a cylinder, that is, a one-dimensional problem is considered [17].

2. Since the water is in excess during the delignification, the change in its concentration during the reaction is neglected, and the delignification reaction is considered as a monomolecular, or first-order reaction [18].

3. The material after steam explosion treatment, subjected to delignification, consists of cellulose, low molecular weight lignin and part of natural lignin that has not undergone any changes during steam explosion treatment; hemicelluloses are absent [19-22].

4. The gas and solid phases are in a state of thermodynamic equilibrium, so the temperature of the particle is determined by the temperature of the resulting vapor-gas mixture.

3. Mathematical modeling of the delignification process of activated wood

After the process of steam explosion activation of wood, the raw material is a fibrous intermediate product with a component composition directly depending on the conditions of steam explosion treatment:

- alpha cellulose:

$$C_1 = 88.53\exp(-0.0016t) - 0.08t^{1.56}\exp(-0.01t);\qquad (1)$$

- lignin

$$C_2 = 10.89\exp(0.011t) - 6.43\cdot10^{-6}t^{3.1};\qquad (2)$$

- hemicellulose

$$C_3 = 7.65\exp(0.012t) - 4.68\cdot10^{-9}t^{4.7}\exp(-0.008t);\qquad (3)$$

- extractives

$$C_4 = 120.92\exp(-0.014t) - 0.25t^{1.5}\exp(-0.025t);\qquad (4)$$

- total solid products from spruce

$$C = 78.59\exp(0.0026t) - 2.7\cdot10^{-16}t^{8.7}\exp(-0.03t).\qquad (5)$$

The resulting fibrous product is subjected to washing from the degradation products of hemicelluloses and a portion of the low molecular weight lignin in a drum vacuum filter. The rate of filtration of the washing liquid (1/s) is determined from the Poiseuille equation [22]:

$$q = \frac{\Delta p}{Rl},\qquad (6)$$

where Δp - pressure drop on the sides of the layer of lignocellulosic material, Pa; l - layer thickness, m; R - the coefficient of filter resistance, equal to [27]

$$R = \frac{32\mu}{\varphi d^2},\qquad (7)$$

where μ - is the dynamic viscosity of the liquid, Pa·s; φ - free area ratio of the filter; d - is the average diameter of the capillaries in the layer, m.

The degree of selection, or washing efficiency, is the ratio of the amount of dissolved substances in the obtained solution G to their quantity in the initial solution after the steam explosion treatment G_0 [23]:

$$\eta = \frac{G}{G_0}.\qquad (8)$$

G_0 is the mass of substances formed as a result of steam explosion treatment (sugars and low molecular weight lignin) and subject to extraction during washing [24]:

$$G_0 = m_{a.d.w.}\cdot(C_{3_0} + C_{2_0} - C_3 - C_2),\qquad (9)$$

where $m_{a.d.w.}$ is the mass of absolutely dry wood exposed to steam explosion treatment, C_{3_0}, C_{2_0} are the mass fractions (% of the a.d.w.) of hemicelluloses and lignin in the original wood before the steam explosion treatment, C_3, C_2 - mass fractions (% of the a.d.w.) of hemicelluloses and lignin after steam explosion treatment in a material subjected to washing.

The washed fiber is loaded into the cooker, where the cooking liquor is circulated by the pump and the system heats up. Diffusive impregnation is described according to the Fick law, in which the concentration gradient (g/s) is defined as:

$$\frac{dG}{d\tau} = -DA_{sa}m_{lc}\frac{dC}{dl},\qquad (10)$$

where G is the amount of diffusible cooking liquor, g; τ - duration of the diffusion process, s; m_{lc} - the mass of absolutely dry lignocellulosic fiber, which is supplied for delignification; C - concentration of solute, g/cm^3; l is the distance at which diffusion takes place (mean particle diameter), cm; D - diffusion coefficient, depending on the parameters of the solvent, penetrating particles and temperature, cm^2/s.

$$m_{lc} = m_{a.d.w.} \cdot (C_1 + C_2 + C_3 + C_4), \tag{11}$$

where C_1, C_2, C_3, C_4 - the concentration of cellulose, lignin, hemicelluloses, extractives respectively after washing, determined by the expressions (1 – 4).

The specific surface area A_{sa} of the fibrous lignocellulosic material formed after steam explosion treatment (m^2/kg) is determined from the data of:

$$A_{sa} = \frac{0.01 - 0.004 p_{set} - 0.0003 p_{set}^2 + 0.0003 p_{set}^3 + 0.0008 \ln \tau_{dec} + 0.0015 (\ln \tau_{dec})^2}{1 - 0.66 p_{set} + 0.12 p_{set}^2 + 0.144 \ln \tau_{dec} + 0.087 (\ln \tau_{dec})^2} \tag{12}$$

at the set temperature t_{set}, the pressure p_{set} of steam explosion treatment and the decompression time of the reactor τ_{dec}, where p_{set} is also determined by the empirical expression [25-26]

$$p_{set} = 2.04 e^{2.7 \cdot t_{set}^{0.3}}. \tag{13}$$

The rate of direct reaction (change in concentration in time) at any time is directly proportional to the production of the volume concentrations of substances participating in the reaction [28-29]:

$$\frac{dC_i}{d\tau} = k_m \prod_{i=1}^{n} C_i^{\chi_i}, \tag{14}$$

where k_m is the mass transfer coefficient complicated by chemical transformation, which can be determined from the expressions obtained as a result of modeling in [30]

$$k_m = 5736.4 \exp(-0.0016 \tau^{0.2}) - 5978.8 \tau^{0.01} \exp(-0.056 \tau^{0.1}). \tag{15}$$

The weight and volume concentrations are connected by the relation [31]

$$\frac{dC_i}{d\tau} = \frac{\rho_{mix}}{\mu_i} \frac{d\chi_{i_{ch.r}}}{d\tau} + \frac{\overline{\chi}_i}{\mu_i} \frac{d\rho_{mix}}{d\tau}. \tag{16}$$

During the passage of the cooking liquor through the layer of activated wood, due to heat-mass transfer and chemical reactions, the content of delignification products in the cooking liquor changes with time and the temperature over the cross section of the particle is determined from the heat and material balances recorded in differential form. The solid phase is represented by lignin and cellulose [32-33], the liquid phase is water contained in the fiber after washing, the reagent solution, the gas phase is air in the cooker, the gases formed during the interaction of lignin with the reagent (methyl mercaptan, dimethyl sulfide, hydrogen sulphide, turpentine, etc.), water and organic acids vapors. Since the process is formalized, the steam explosive activated material after washing is considered as lignocellulose - a mixture consisting of two main constituents: cellulose and lignin with known mass fractions.

The change cellulose weight is insignificant, and it is not taken into account in the description, that is $\partial m_c = 0$.

The change in lignin mass will be presented as (cellulose does not participate in the delignification process):

$$\frac{dm_l}{d\tau} = -k_l \cdot m_{a.d.w.} \cdot C_2, \tag{17}$$

where m_l - is lignin mass, τ - is delignification time, k_l - is rate constant of delignification.

The reaction rate constant for lignin k_l (1/h) is calculated by the formula [34-35]:

$$k_l = \frac{1}{\tau} \ln \frac{m_{lc}}{m_{lc} - x}, \tag{18}$$

where m_{lc} - the amount of initial lignocellulose or lignin; $(m_{lc} - x)$ - the amount of residue (yield of cellulose) or lignin after τ delignification hours. The weight of technical pulp after delignification is determined by the formula:

$$m_{lc} = m_{a.d.w.} \cdot (C_2 + C_1). \tag{19}$$

Steam and gas formation during delignification is accompanied by absorption and release of heat, which in turn has a significant effect on the temperature change of both the individual particle and the layer as a whole. The temperature of the gas stream will determine the temperature of the solid particle, and the heat balance equation will be written as:

$$(c_c \cdot m_c + c_l \cdot m_l + c_{sg} \cdot m_{sg})\frac{\partial T}{\partial \tau} = \frac{\partial T}{\partial l}(\lambda_p \cdot \frac{\partial T}{\partial l}) - c_{sg} \cdot m_{sg} \cdot w_{sg}\frac{\partial T}{\partial l} + q_{ch.r.}, \quad (20)$$

where λ_p is the coefficient of thermal conductivity of a lignocellulosic particle, $q_{ch.r.}$ is the heat energy released (absorbed) by a chemical reaction; m - is the mass of matter per unit volume of the cooker.

The heat balance [36-37] equation for the liquid phase in the differential form is written as:

$$c_{mix}m_{mix}dT + c_{mix}dm_{mix}T + dc_{mix}m_{mix}T = \sum_{i=1}^{n} Q_{iload}c_iT_{iH}d\tau - \sum_{i=1}^{n} j_ic_{li}TFd\tau - \sum_{i=1}^{n} j_ir_iFd\tau + \sum_{i=1}^{n} q_{i_{ch.r.}}\varpi_i\rho_{li}Vd\tau . \quad (21)$$

The change in the heat content of a mixture of liquid components in the cooker is determined from the left side of equation (21). The first term on the right side determines the amount of heat that came with the components at boot time; the second term characterizes the loss of heat carried away by the components of the vapor mixture; the third term determines the heat consumption for the formation of vapors; the fourth is the heat of the chemical reaction. The material balance in the differential form will be written in the form:

$$dm_{mix} = \sum_{i=1}^{n} Q_{iload}d\tau - \sum_{i=1}^{n} j_iFd\tau - \sum_{i=1}^{n} \varpi_i\rho_{li}Vd\tau . \quad (22)$$

The change in mass of the mixture in the cooker is determined from the left side of equation (22). The first term on the right side allows one to estimate the mass of the loaded components, the second allows for evaporation of the components, the third determines the change in mass due to chemical reactions.

The change in the heat capacity of a liquid mixture is defined as the sum of the products of the heat capacity of the i-th component of the mixture at its concentration:

$$dc_{mix} = \sum_{i=1}^{n} c_{li}d\chi_i . \quad (23)$$

The concentration of the i-th component changes as a result of evaporation [38], the course of the chemical reaction and due to the loading:

$$d\chi_i = d\chi_{ievap} + d\chi_{ich.r.} + d\chi_{iload} . \quad (24)$$

The change in the concentration of components in real cooking liquids can be determined as:

$$d\chi_{ievap} = \frac{dP_i}{\gamma_i \exp(A_i - \frac{B_i}{T})} - \frac{\chi_i B_i}{T^2}dT - \frac{\chi_i}{\gamma_i}d\gamma_i . \quad (25)$$

The change in the concentration of the mixture component per unit time determines the rate of the chemical reaction:

$$d\chi_{ich.r.} = \frac{\mu_i}{\rho_{mix}}\left[k\prod_{i=1}^{n}\left(\frac{\rho_{mix}\chi_i}{\mu_i}\right)^{\chi_i}\right]d\tau - \frac{\chi_i}{\rho_{mix}}d\rho_{mix} . \quad (26)$$

The change in the concentration of the component due to the loading into the cooker is determined by the relation

$$d\chi_{iload} = \frac{Q_i\rho_i}{m_{mix}}d\tau . \quad (27)$$

IOP Conf. Series: Materials Science and Engineering **221** (2017) 012009 doi:10.1088/1757-899X/221/1/012009

Assuming that the components of the gas-vapor mixture are perfectly mixed, we will write the equation of the material balance of the gas-air mixture in a differential form for the stage of unloading the cooker due to a pressure drop:

$$\frac{dm_{mix}}{d\tau} = \frac{dm_{vap}}{d\tau} + \frac{dm_{g-air}}{d\tau} = \frac{d\rho_{mix}}{d\tau} V_{free} = \frac{d\rho_{g-air}}{d\tau} V_{free} + \frac{d\rho_{vap}}{d\tau} V_{free}. \quad (28)$$

The temperature of the vapor-gas-air mixture is determined from the heat balance equation for the vapor-gas-air mixture:

$$\frac{dT_1}{d\tau} = \frac{\alpha(T_l - T_g)F_{load}}{c_l m_l + c_g m_g} + \frac{j_g F_{load} c_g T_l}{c_l m_l + c_{air} m_{air}} + \frac{j_{vap} F_{load} c_{vap} T_l}{c_l m_l + c_{air} m_{air}} + \frac{j_{air} F_2 c_{air} T_{air}}{c_l m_l + c_{air} m_{air}} - \frac{Q_{vap} \rho_{vap} c_{vap} T_g}{c_l m_l + c_{air} m_{air}} - \frac{Q_{g-air} \rho_{g-air} c_{g-air} T_g}{c_l m_l + c_{air} m_{air}}. \quad (29)$$

The density of the components is determined by the equation

$$\frac{d\rho_{vap-g-air}}{d\tau} = \sum_{j=1}^{m} \frac{dy_j}{d\tau} \frac{\mu_j p_j}{RT} + \sum_{j=1}^{m} \frac{dp_j}{d\tau} \frac{\mu_j y_j}{RT} - \sum_{j=1}^{m} \frac{dT}{d\tau} \frac{\mu_j y_j p_j}{RT^2} + \sum_{i=1}^{n} \frac{dy_i}{d\tau} \frac{\mu_i p_i}{RT} +$$
$$+ \sum_{i=1}^{n} \frac{dp_i}{d\tau} \frac{\mu_i y_i}{RT} + \sum_{i=1}^{n} \frac{dT}{d\tau} \frac{\mu_i y_i p_i}{RT^2} + \frac{dy_{air}}{d\tau} \rho_{vap-g-air} \quad (30)$$

The heat capacity of the above mixtures can be determined according to the additivity rule [39-40].

$$c_{vap-g-air} = \frac{c_{air} p_{air} + \sum_{i=1}^{n} c_i p_i + \sum_{j=1}^{n} c_j p_j}{p_{air} + \sum_{i=1}^{n} p_i + \sum_{j=1}^{n} p_j}. \quad (31)$$

4. Method of experimental research

In the course of the work we used wood sorted from the species of coniferous species sorted to the middle size 5x5x0.5 mm. Samples of sawdust were subjected to chemical analysis for the quantitative content of extractives, holocellulose, alpha-cellulose, lignin, easily hydrolyzable and hardly hydrolyzable polysaccharides by known methods. The mass fractions of the components (in % of the a.d.) of pine were: 51 cellulose; lignin - 24.1; pentosans - 10.2; hexosans-7.6; extractive substances – 6.7. For a given wood species, a number of experiments on steam explosion treatment were carried out. The temperature of steam explosion treatment was: 130, 160, 190, 220 °C, the processing time was 5, 10, 15, 20 minutes. In all the samples of the blasted and washed lignocellulosic material, the specific surface area was determined by the BET method. The analysis data were used to calculate the equations of the mathematical model.

To carry out experimental studies of the influence of temperature on the rate constant of delignification of coniferous wood preliminarily activated by steam explosion treatment, the samples were subjected to continuous delignification with sulphate and sulphite liquor at the plant, the scheme of which is shown in Fig. 2.

The plant for delignification of activated wood works as follows. The raw material in the form of wood chips is loaded into the feed hopper 1, where it is saturated with water, and then enters the steam explosion reactor 2, where saturated water vapor is fed. After the raw material has been left in the reactor, a relief valve 3 opens from the drive 7 and the activated lignocellulosic material is expulsed into the cooker 4. The discharge valve closes again, the pump 10 is turned on, the liquor is pumped into the cooker, which circulates through the heat exchanger 5 and is heated through the wall by steam coming from superheater of the steam generator. At the end of the cooking process, the steam mixture is blown off, the valve 9 opens, and the pulp is blown into the blowing tank 8, from where it enters the receiver 11. The unit is equipped with an automation system. So, in there are temperature sensors T1 and pressure P1 in the reactor for steam explosion treatment, there are two temperature sensors T3 and T4, as well as a pressure sensor P2 in the cooker. To control the temperature and flow rate of the circulating cooking liquor, the sensors T5 and G2 are installed at the pump outlet.

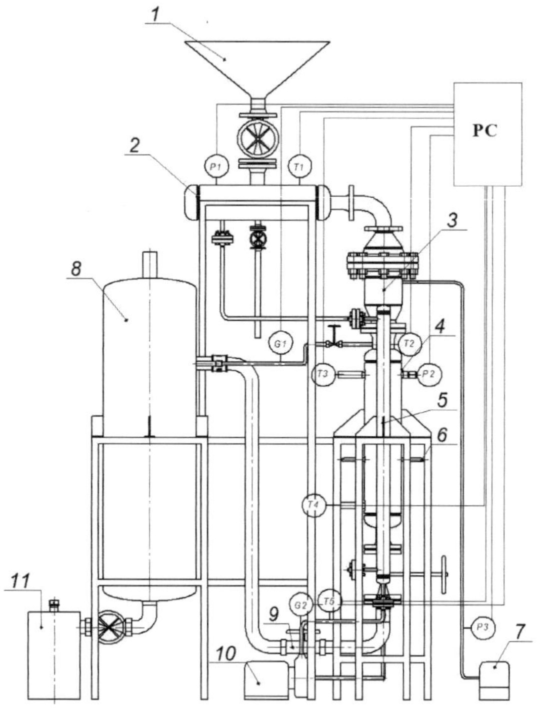

Figure 2. Scheme of installation for wood delignification, previously activated by steam explosion treatment: 1 - bunker for loading wood material, 2 - reactor for steam-explosive activation, 3 - discharge unit, 4 - cooker, 5 - heat exchanger, 6 - thermocouples, 7 - liquid pump, 8 - blowing tank, 9 - blowing valve, 10 - circulating pump, 11 - product receiver

To regulate the temperature of the steam in the heat exchanger, a thermocouple T2 is installed. The pressure in the discharge line of the discharge unit is controlled by the pressure sensor P3. The gas flow rate at the primary blowing is controlled by the flowmeter G1 of regulation of the temperature of pressure and flow. All the sensor readings are collected in the controller and converted into an analog-digital converter (PC). The temperature in the cooker is controlled by the temperature of the steam in the heat exchanger, which is formed in the steam generator. The cooking temperature was 165-170° C, the duration - up to 5 hours. For comparison, similar experiments were conducted on cooking chips, used in traditional processes for the production of cellulose.

5. Results of experiments and discussion

To assess the efficiency of the steam explosion treatment, a number of sulphate pulpings were carried out on the experimental setup of both activated lignocellulosic material and conventional pine chips before delignification. The continuously circulating liquor was heated in a heat exchanger to a cooking temperature of 165 ° C, the concentration of the active alkali was 49 g/l, the cooking modulus was 4. The primary lignin content of the chips was 18% of the a. d. for activated lignocellulosic material (ALM) – 14.1% of a. c. (processed by steam explosion treatment at a temperature of 220 °C for 5 minutes, A_{sa} = 6.03 m^2/g). The comparative results of the two pulpings are shown in Fig. 3.

A comparative analysis of the results of two pulpings showed that when cooking activated lignocellulosic material, the temperature increase occurs faster than in the case of chips, which is primarily due to the high moisture state after steam explosion treatment, which facilitates accelerated impregnation of liquor into the interior of the particles, hence the lignin destruction reaction begins earlier, than in the case of chips. This is proved by the data of curves 4 and 5, from which it can be seen that the transition of lignin to alkali from the activated material occurs much faster than in the case of chips, and the delignification process is about 1.5 - 2 times faster. This is also facilitated by a lower initial content of lignin at the beginning of delignification, in view of its partial degradation in steam explosion treatment. The yield of cellulose remains the same in both cases, however, when delignification of activated wood occurs earlier at the same time of cooking the chips and ALM, an increase in the yield of cellulose is likely.

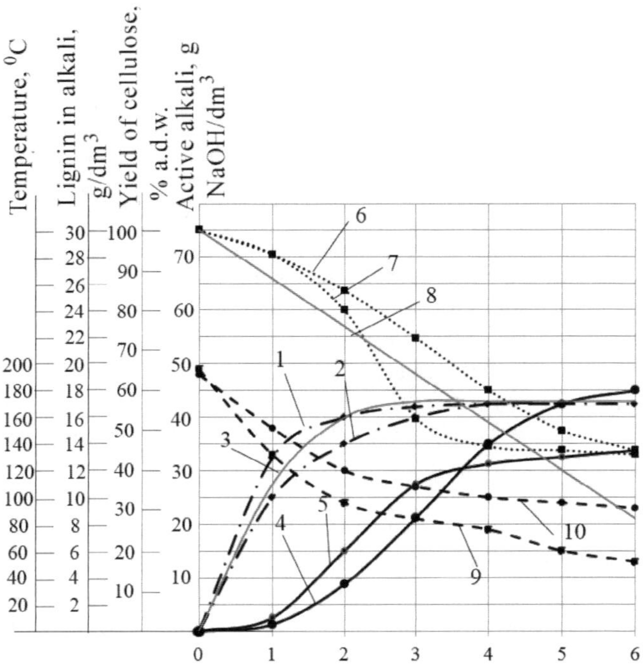

Figure 3. Change in the composition of solid residues and liquors during sulphate pulping: 1- temperature (ALM), 2 - temperature (chips), 3 - calculated temperature curve (for ALM), 4 - lignin in alkali (chips), 5 - lignin in alkali (ALM), 6 - yield of cellulose (chips) 7 - yield of cellulose (ALM), 8 - calculated cellulose yield curve (for ALM), 9 - consumption of active alkali (chips), 10 - consumption of active alkali (ALM).

Curves 9, 10 clearly illustrate that the consumption of active alkali with an equal initial concentration in both cases is quite different. For the same period of cooking, alkali is consumed in the case of activated lignocellulose by 15-17% less than for cooking chips. All these data prove the high efficiency of the application of steam-explosive activation of wood raw materials before delignification. A key role in improving the efficiency of the delignification process is undoubtedly played by the process of explosive separation into fibers, defibrillation at high temperatures of steam-explosive activation. Steam treatment (steam prehydrolysis) allows only to reduce the amount of lignin and hemicelluloses in the original wood, which is also important, and to create conditions for explosive boiling up of superheated liquid inside the material. The calculated curves 3 and 8 make it possible to confirm the correctness of the conclusions in the simulation, since the discrepancy with the experimental data is 6.4% and 13.1%, respectively. An increased percentage of the discrepancy with the experiments to determine the yield of cellulose is acceptable, since the very definition of the mass fraction of cellulose in a solid residue implies a series of experiments, when summing up the errors of which, such a high error percentage is formed.

The increase in the efficiency of delignification is caused by the highly developed specific surface of the material after steam-explosive activation and the increase in the size of the pores and capillaries, the change in the structure of the cell wall of the wood fiber, the high moisture content of the material after washing, the low initial content of lignin at the beginning of delignification, in view of its partial degradation during steam explosion treatment and almost complete absence of hemicelluloses. The use of explosive separation of the activated material on the particles doesn't only give a high specific surface, but also mechanically affects the condensed lignin fractions formed in the process of high-temperature steam hydrolysis. These insoluble lignin fractions are destabilized and undergo degradation, which subsequently facilitates both washing and diffusion impregnation and delignification itself.

The optimal parameters for steam explosion treatment are: type of wood - spruce, pine; fractional composition of raw materials - flake, chips; the temperature of the steam is from 210 to 230 °C; soaking time at the set temperature - 10 minutes; pressure - from 2.2 to 2.9 MPa; decompression time - 0.2 s; extractant of reducing substances (sugars) - ethanol-water solution with a temperature of 90 ° C;

extractant of low molecular weight lignin - sodium hydroxide from 0.4 to 2% with a temperature of 20 ° C.

The positive effect of the preliminary steam-explosive activation of wood before delignification consists in increasing the productivity of the cooker, reducing the consumption of reagents for cooking, the possibility of varying the reagents, and also in obtaining dissolved hemicelluloses during the steam explosion treatment. The shown efficiency of the process is fully confirmed by experiments and is in good agreement with the calculated data with an average error of 7.5%, which indicates the correctness of acceptance of the assumptions about the process, the adequacy of the developed mathematical model to the actual process of wood delignification, previously activated by steam explosion treatment, and the correctness of the assumptions. Mathematical modeling, as well as experimental research, is a source of data on the process of delignification of wood, previously activated by steam explosion treatment, and allows to determine the influence of various parameters on the efficiency of the delignification process. The constructed mathematical model was tested for adequacy by known methods of mathematical statistics. Under identical conditions, the results of experimental studies were compared with theoretical calculations.

6. Conclusion

Perfection of the mechanism of wood delignification process, previously activated by steam explosion treatment, consists in the development of new calculation methods aimed at increasing the efficiency of the process, and the improvement of technology consists in the use of new methods of processing wood raw materials, on the basis of which the problem of increasing the stability of positioning of domestic technologies in the world market of wood chemical Industry, particularly in the pulp and paper sector, is solved.

Acknowledgments

The submissions were received during the execution of the state task within the framework of the "Initiative scientific projects" event on the topic No. 13.5443.2017 / BC "Modification of the physical and chemical properties of wood biomass influencing the performance characteristics of the created materials".

References

[1] Gujjala, L. K., Bandyopadhyay, T. K., & Banerjee, R. 2016 Kinetic modelling of laccase mediated delignification of Lantana camara. Bioresource technology, 212: 47-54.

[2] Dagnino, E. P., Felissia, F. E., Chamorro, E., & Area, M. C. 2017 Optimization of the soda-ethanol delignification stage for a rice husk biorefinery. Industrial Crops and Products, 97: 156-165.

[3] Dutta, S. K., Halder, G., & Mandal, M. K. 2014 Modeling and optimization of bi-directional delignification of rice straw for production of bio-fuel feedstock using central composite design approach. Energy, 71: 579-587.

[4] Subhedar, P. B., Ray, P., & Gogate, P. R. 2017 Intensification of delignification and subsequent hydrolysis for the fermentable sugar production from lignocellulosic biomass using ultrasonic irradiation. Ultrasonics Sonochemistry.

[5] Qing, Q., Zhou, L., Guo, Q., Huang, M., He, Y., Wang, L., & Zhang, Y. 2016 A combined sodium phosphate and sodium sulfide pretreatment for enhanced enzymatic digestibility and delignification of corn stover. Bioresource Technology, 218: 209-216.

[6] Karimi, M., Esfandiar, R., & Biria, D. 2017 Simultaneous delignification and saccharification of rice straw as a lignocellulosic biomass by immobilized Thrichoderma viride sp. to enhance enzymatic sugar production. Renewable Energy, 104: 88-95.

[7] Prathyusha, N., Kamesh, R., Rani, K. Y., Sumana, C., Sridhar, S., Prakasham, R. S., & Kumar, M. P. 2016 Modelling of pretreatment and saccharification with different feedstocks and kinetic modeling of sorghum saccharification. Bioresource technology, 221: 550-559.

[8] Susilo, J., & Bennington, C. P. J. 2007. Modelling kappa number and pulp viscosity in industrial oxygen delignification systems. Chemical Engineering Research and Design, 85(6): 872-881.

[9] Salmi, T., Wärn, J., Mikkola, J. P., & Rönnholm, M. 2005. Modelling and simulation of porous, reactive particles in liquids: delignification of wood. Computer Aided Chemical Engineering, 20: 325-330.

[10] Vega, A., Bao, M., & Lamas, J. 1997. Application of factorial design to the modelling of organosolv delignification of Miscanthus sinensis (elephant grass) with phenol and dilute acid solutions. Bioresource technology, 61(1): 1-7.

[11] Kumar, R., Hu, F., Hubbell, C. A., Ragauskas, A. J., & Wyman, C. E. 2013 Comparison of laboratory delignification methods, their selectivity, and impacts on physiochemical characteristics of cellulosic biomass. Bioresource technology, 130: 372-381.

[12] Rocha, G. J. M., Gonçalves, A. R., Oliveira, B. R., Olivares, E. G., & Rossell, C. E. V. 2012 Steam explosion pretreatment reproduction and alkaline delignification reactions performed on a pilot scale with sugarcane bagasse for bioethanol production. Industrial Crops and Products, 35(1): 274-279.

[13] Marton, G., Kovacs, M., Oreopoulou, V., & Koukios, E. G. 1988 Modelling of biomass fractionation by prehydrolysis-delignification. Chemical engineering science, 43(8): 1807-1812.

[14] Huron, M., Hudebine, D., Ferreira, N. L., & Lachenal, D. 2016 Impact of delignification on the morphology and the reactivity of steam exploded wheat straw. Industrial Crops and Products, 79: 104-109.

[15] Hubbell, C. A., & Ragauskas, A. J. 2010 Effect of acid-chlorite delignification on cellulose degree of polymerization. Bioresource Technology, 101(19): 7410-7415.

[16] Pinto, P. C., Oliveira, C., Costa, C. A., Gaspar, A., Faria, T., Ataíde, J., & Rodrigues, A. E. 2015 Kraft delignification of energy crops in view of pulp production and lignin valorization. Industrial Crops and Products, 71: 153-162.

[17] Ratti, R. P., Delforno, T. P., Sakamoto, I. K., & Varesche, M. B. A. 2015 Thermophilic hydrogen production from sugarcane bagasse pretreated by steam explosion and alkaline delignification. International Journal of Hydrogen Energy, 40(19): 6296-6306.

[18] Timilsena, Y. P., Abeywickrama, C. J., Rakshit, S. K., & Brosse, N. 2013 Effect of different pretreatments on delignification pattern and enzymatic hydrolysability of miscanthus, oil palm biomass and typha grass. Bioresource technology, 135: 82-88.

[19] Marton, G., Kovacs, M., Oreopoulou, V., & Koukios, E. G. 1988 Modelling of biomass fractionation by prehydrolysis-delignification. Chemical engineering science, 43(8): 1807-1812.

[20] Mafuleka, S., & Kana, E. G. 2015 Modelling and optimization of xylose and glucose production from napier grass using hybrid pre-treatment techniques. Biomass and Bioenergy, 77: 200-208.

[21] Jabbari, M., et al. 2016 Drying of a tape-cast layer: Numerical modelling of the evaporation process in a graded/layered material. International Journal of Heat and Mass Transfer, 103: 1144-1154. DOI:10.1016/j.ijheatmasstransfer.2016.08.073

[22] Fomin A.A., Gusev V.G., Yudin R.V., Timerbaev N.F. and Retyunskiy O.Yu. 2016 Mechanical treatment of raw waste lumber an effective way to preserve the ecology and resources. IOP Conference Series: Materials Science and Engineering 142(1): 012091.

[23] Timerbaev N.F., Safin R.G., Ziatdinova D.F., Fomin A.A. and Mokhovikov A.A. 2016 The development of experimental setups and experimental studies of the process of energy-technological processing of wood. IOP Conference Series: Materials Science and Engineering 142(1): 012096. DOI:10.1088/1757-899X/142/1/012096

[24] Prosvirnikov D.B., Ziatdinova D.F., Timerbaev N.F., Saldaev V.A. and Gilfanov K.H. 2016 Mathematical modelling of the steam explosion treatment process for pre-impregnated lignocellulosic material. IOP Conference Series: Materials Science and Engineering 124(1): 012087. DOI:10.1088/1757-899X/124/1/012087

[25] Safin R., Barcík Š., Shaikhutdinova A., Safina A., Kaynov P. and Razumov E. 2015 Development of the energy-saving technology of thermal modification of wood in saturated

steam. Acta Facultatis Xylologiae 57(2): 39-47. DOI:10.17423/afx.2015.57.2.04
[26] Lashkov V.A., Levashko E.I. and Safin R.G. 2001 Heating of production wood chips in a saturated-steam medium. Inzhenerno-Fizicheskii Zhurnal, 74(1): 80-83.
[27] Ivanov, E. V., M. V. Shvyrev, and M. A. Artemova. 2004 Filtration of the extracting agent in porous particles with entrapped gas at low-frequency pressure oscillations in the extractor. Russian journal of applied chemistry, 77(10): 1662-1666.
[28] Bouzid, Majda, et al. 2011 In-pore tensile stress by drying-induced capillary bridges inside porous materials. Journal of colloid and interface science, 355(2): 494-502.
[29] Defraeye, Thijs, Bert Blocken, and Jan Carmeliet. 2013 Influence of uncertainty in heat–moisture transport properties on convective drying of porous materials by numerical modelling. Chemical Engineering Research and Design, 91(1): 36-42.
[30] Prosvirnikov D.B., Safin R.G., Ziatdinova D.F., Timerbaev N.F., and Lashkov V.A. 2016 Multifactorial modelling of high-temperature treatment of timber in the saturated water steam medium. IOP Conference Series: Materials Science and Engineering, 124(1): 012088. DOI:10.1088/1757-899X/124/1/012088
[31] Van Belleghem, Marnix, et al. 2014 Validation of a coupled heat, vapour and liquid moisture transport model for porous materials implemented in CFD." Building and Environment, 81: 340-353. DOI:10.1016/j.buildenv.2014.06.024
[32] Sadrtdinov A.R., Sattarova Z.G., Prosvirnikov D.B. and Tuntsev D.V. Modeling of thermal treatment of wood waste in the gasifiers. Proceedings of 2015 International Conference on Mechanical Engineering, Automation and Control Systems, MEACS 2015. art. no. 7414914. DOI:10.1109/MEACS.2015.
[33] Timerbaev N.F., Safin R.R., Safin R.G. and Ziatdinova D.F. 2014 Modeling of the process of energy-technological treatment of wood waste by method of direct-flow gasification. Journal of Engineering and Applied Sciences (EAAS) 9(5): 141-146
[34] Zhang, Hu, et al. 2014 Experimental study on the kinetics of water vapor sorption on the inner surface of silica nano-porous materials. International Journal of Heat and Mass Transfer, 78: 947-959. DOI:10.1016/j.ijheatmasstransfer.2014.07.047
[35] Choudhary, Anupama, Devendra Kumar, and Jagdev Singh. 2015 A fractional model of fluid flow through porous media with mean capillary pressure. Journal of the Association of Arab Universities for Basic and Applied Sciences, 21: 59-63. DOI:10.1016/j.jaubas.2015.01.002
[36] Sadrtdinov A.R., Safin R.G., Gerasimov M.K., Petrov V.I., and Gilfanov K.K. 2016 The mathematical description of the gasification process of woody biomass in installations with a plasma heat source for producing synthesis gas. IOP Conference Series: Materials Science and Engineering, 124(1): 012092. DOI:10.1088/1757-899X/124/1/012092
[37] Sychevskii, V.A. 2015 Drying of colloidal capillary-porous materials. International Journal of Heat and Mass Transfer, 85: 740-749.
[38] Lashkov V.A., Sattarova Z.G., Taymarov M.A., Gerasimov M.K. and Halitov R.A. 2016 Modeling of a reduction zone of the gasifier installation. IOP Conference Series: Materials Science and Engineering, 124(1): 012111. DOI:10.1088/1757-899X/124/1/012111
[39] Sadrtdinov A.R., Esmagilova L.M., Saldaev V.A., Sattarova Z.G. and Mokhovikov A.A. 2016 Mathematical modeling for the development of equipment for thermochemical processing of wood waste in to dimethyl ether. IOP Conference Series: Materials Science and Engineering, 142(1): 012094. DOI:10.1088/1757-899X/142/1/012094
[40] Yang, Lianfa, Haisong Rong, and Yulin He. 2014 Deformation behavior of a thin-walled tube in hydroforming with radial crushing under pulsating hydraulic pressure. Journal of materials engineering and performance 23(2): 429-438. DOI:10.1007/s11665-013-0783-9

IOP Conf. Series: Materials Science and Engineering 221 (2017) 012010 doi:10.1088/1757-899X/221/1/012010

Mechanization of Continuous Production of Powdered Cellulose Technology

D B Prosvirnikov, R G Safin, I R Akhmetshin, M A Taimarov, N F Timerbaev

Institute of petroleum and chemical engineering, Kazan national research technological University, Russian Federation, Republic of Tatarstan, Kazan, K. Marx street 68

E-mail: prosvirnikov_dmi@mail.ru

Abstract. The article presents the mechanization of the process for obtaining powdered cellulose using the technology of steam explosion treatment of lignocellulosic material. The presented unit combines the methods of steam explosion treatment and acid hydrolysis of cellulose-containing raw materials.

1. Introduction

Powder cellulose is a finely dispersed product of the hydrolytic destruction of cellulose, most fully released from the cell wall of vascular plants by the crystalline part of the cellulose chain. The structure of cellulose chains itself combined into a bundle of microfibrils is an amorphous and crystalline areas interleaved along the chain direction. In crystalline sections of cellulose there is a crystallographic orientation of macromolecules and a strong intermolecular interaction. In the amorphous regions, however, only the overall longitudinal orientation of the chains is retained and there is no strict order of the location of the chain sections [4]. Thus, when exposed to cellulose, amorphous areas are first destroyed, and then crystalline areas. In the complete hydrolytic destruction of cellulose, a breakdown of glycosidic bonds occurs and the product of complete destruction of cellulose is glucose [2]. Thus, the essence of the methods for obtaining powdered cellulose is a strictly controlled incomplete treatment of cellulosic material.

2. The background for using lignocellulosic raw materials for the production of powdered cellulose

For the first time microcrystalline cellulose was accidentally discovered in 1955 by Dr. O.A. Battista and Patricia Smith in the laboratory, while they were trying to create a strong viscose cord for automobile tires from cellulose. The discovered features of the structure and properties of the new material, as well as the possibilities of its application, were used as the reason for the organization of its industrial production [1, 3]. Thus, powdered cellulose became of practical interest.

A characteristic feature of powdered celluloses, and in particular of microcrystalline cellulose, is the ability to form gel-like dispersions in water. Due to the ability to be dispersed in a water medium, microcrystalline cellulose has a highly developed hydrophilic surface. Gels of MCC have a high ability to retain water. Since powdered celluloses have a large specific surface area, therefore surface activity increases, they show high sorption properties.

The complexity of the processes for the production of powdered cellulose, when natural resources are used as feedstock, lies in the fact that such raw materials apart the cellulose itself contain a significant number of foreign compounds. These include lignin, hemicelluloses, various extractives and a small percentage of mineral substances [11]. For the production of powdered cellulose from plant cellulose-containing raw materials, it is first necessary to isolate the cellulose itself. Further operations are to hydrolyze the isolated cellulose, which leads to the destruction of glycosidic bonds in amorphous sections of the chains. Thus, the process of production of powdered cellulose from the original plant material consists of two essentially different stages of impact on the organic material. Particular attention should be paid to the fact that the productivity of the subsequent stage of hydrolytic degradation depends on the initial stage of cellulose evolution [6]. After all, the use of not fully isolated cellulose containing inclusions of other components of plant raw materials significantly worsens the effect of subsequent steps aimed at directly affecting the cellulose chains themselves. Also, the complexity of obtaining powdered cellulose on an industrial scale is to integrate most of the operations into a combined production stream. This is due to the fact that, as a rule, traditional methods of isolation of cellulose have a long duration in time, which considerably complicates the organization of a continuous flow. One promising way of pretreating wood pulp with the subsequent production of cellulose is the steam explosion method, which allows processing almost any cellulose-containing raw material [7, 9].

To date, more interesting are the methods for obtaining powdered celluloses from various plant raw materials with a high content of cellulose. For example, in Brazil, research is focused on the use of waste from various industries, from fruit plantations to the recycling of cellulose materials.

According to the mentioned above, it is of practical interest to provide optimized methods for the industrial production of powdered celluloses that, in addition to integration and combining all operations for the production of powdered cellulose into one continuous process, allow the processing of not only wood raw materials.

An important aspect is that the process of isolating cellulose from plant material is itself laborious and, as a rule, consists of pulping raw materials using environmentally aggressive reagents. In addition, in the production of powdered cellulose, reagents are also used that pose a threat to the environment. All of the above factors require the search and development of such a technological solution that would simplify and modernize the stage of processing of wood raw materials.

To reduce the use of reagents in the production of cellulose from plant raw materials, it is more expedient to apply the method of steam explosion treatment [8, 10]. This method consists in the fact that with a sudden release of pressure, the fibrous raw material, previously saturated with water steam, is destroyed, and a fibrous dispersed material is obtained at the outlet. After this, residual lignin is extracted from the raw material and is further crushed. Further, the cellulose thus obtained is subjected to acid hydrolysis to reduce the degree of polymerization of cellulose. As a rule, concentrated solutions of mineral acids are used for this [5, 13]. The acid concentration also depends on the reaction temperature. The higher the heating temperature, the lower the acid concentration should be used [12].

Thus, the task of mechanization of the technology of continuous production of powdered cellulose is to combine the method of steam explosion, extraction of lignin and acid hydrolysis of the formed cellulose into a single production cycle. The problem is solved by a series connection of a steam explosion reactor, reactors for intermediate operations and an acid hydrolysis reactor. To move the processed material from one stage of the process to the other, a screw conveyor system is chosen.

3. Installation for continuous production of powdered cellulose

An installation for the continuous production of powdered cellulose (Figure 1) comprises a sequentially connected premixing device, a loading and feeding unit, a reactor for steam prehydrolysis of wood raw materials with a transport system, a steam feed and release system, a steam explosion reactor, a cellulose fiber acid hydrolysis reactor with a system for supplying and removing acid, a unit for unloading powdered cellulose.

The premixing device 1 is filled with wood raw material and water or an acid solution is fed through the valve 4. The components are mixed by a stirrer 2 operating from the drive 3. Then the portion of the impregnated raw material from the premixing device is tangentially supplied to the loading and feeding unit 5, which is in the form of two channels with pistons arranged at an acute angle to each other. Such construction allows the raw material to be fed into the plant continuously.

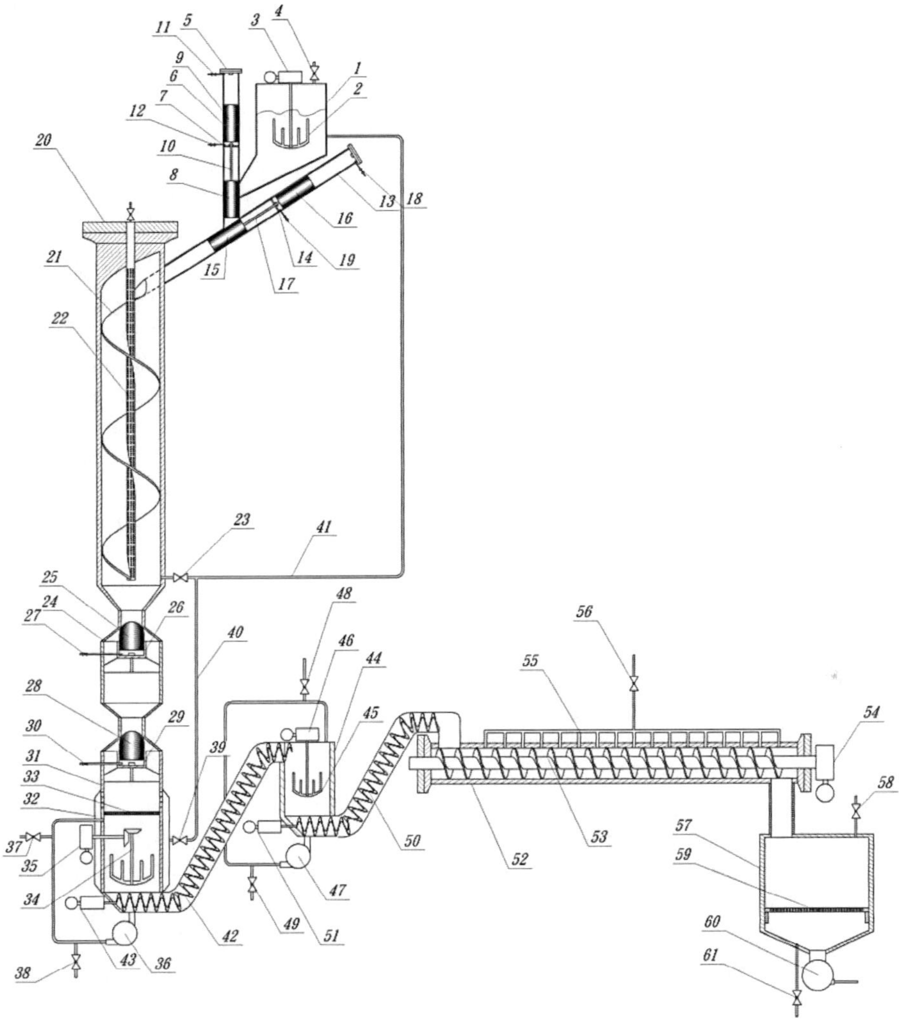

Fig. 1. Layout of installation for continuous production of powdered cellulose

The operation of the loading and feeding unit is arranged so that the pistons move alternately in both channels. When the piston 15 is in the lower position, the piston 8 is in its upper position, thereby the portion of the raw material from the premixing device is fed into the channel 6. When the piston 8 starts to squeeze the raw material, the piston 15 rises. And when it is in its upper position, the feedstock from channel 6 enters channel 13. Then the piston 15 pushes the portion of raw material into the reactor for steam prehydrolysis of wood raw material 20, thereby closing the junction of the two channels. Thus, when the piston 15 is in the lower position, a seal is provided at the inlet of the reactor 20 with a continuous feedstock loading.

The reactor for steam prehydrolysis of wood raw material is located vertically. Inside there is a guiding screw channel. Along the axis of the reactor there is a steam supply and release system made in the form of a coaxial perforated pipe. Thus, moving downward in a vertical reactor, the raw material is uniformly saturated with steam and is subjected to steam hydrolysis throughout the all time in the reactor, as a result of which the hemicellulose of the wood raw material dissolves and the processed lignocellulosic feedstock remains, which accumulates in the lower part of the reactor. The lower part of the reactor is connected to the vapor explosion chamber by a hydraulic valve. When the raw material accumulated in the lower part of the reactor is compacted under the weight of the incoming portions of raw materials, a hydraulic valve opens and the raw material enters the steam explosion chamber. The pressure in the chamber is equalized with the pressure in the reactor, after which the valve closes, thereby ensuring the chamber is sealed when performing a steam explosion treatment.

In the lower part of the chamber 24 there is a hydraulic valve 28 for discharging the processed raw material into the tank for alkaline treatment of lignocellulosic fiber 31. When the lower valve is suddenly opened, the pressure in the steam explosion chamber is instantly released. The superheated liquid in the pre-impregnated lignocellulosic raw material boils over the entire volume of the particle, which leads to separation of the latter into fibers.

Due to the sharp depressurization of the chamber 24, the mass in the form of a lignocellulosic fiber explodes at high speed into an alkaline treatment tank 31 operating at atmospheric pressure. To reduce the speed of the fiber discharged from the chamber, a grate 33 is installed in the middle of the tank 31. The resulting steam is discharged into the tank jacket 32. The alkaline treatment tank allows the extraction of lignin and impurities from the lignocellulosic fiber. In the lower part of the tank 31, the lignocellulosic fiber is mixed with the alkali solution by a stirrer 34 operating from the actuator 35. The formed cellulose fiber by the flexible screw conveyor 42 from the tank 31enters the rinsing device 44 in which the stirrer 45 is mixed with water. Water is constantly removed from the washing zone and a fresh portion of water is fed into it.

From the bottom of the rinsing unit 44, the cellulose fiber is transferred by means of a flexible screw conveyor 50 to an acid hydrolysis reactor 52 operating at atmospheric pressure. The reactor for acid hydrolysis of the cellulosic fiber is arranged horizontally, the nozzles of the reactor acid feed and drain system are distributed along its entire length, and the powdered cellulose discharge unit is equipped with a washing system. The cellulose fiber enters the reactor and is moved within it by a rotating screw 53. The cellulose fiber, being moved in the reactor, is hydrolyzed to the limit of polymerization. The pulp is discharged at the opposite end of the reactor. The mass enters the unloading unit 57, which is equipped with a washing system. The washed powdered cellulose is separated from the wash water by means of a filter 59.

At the output powdered cellulose with a high degree of crystallinity is obtained, which is 0.75-0.85 and a dispersion that is 1.1-1.5 m^2/g. Such powdered cellulose can be used in the pharmaceutical and food industry as an E 460 filler. The food stabilizer E460 is used in the food industry for the process of making food products such as bakery and confectionery, some types of sauces and dairy products, which are low in calories. In addition, often the substance is added to the composition of filter materials for the production of food products. In the pharmacological industry, the use of the food stabilizer E460 is of no small importance. So, this additive can be used to make fillers of certain preparations, sanitary napkins, adhesive dental materials. It is also allowed to have the food stabilizer E460 in the composition of cosmetics, creams and dyes. In addition, the use of the food stabilizer E460 is also known in the chemical industry, where it is used in the process of producing refractory ceramic products, rubber, sorbents, polyurethanes and bitumen heat-resistant coatings

4. Conclusion

The presented installation for the continuous production of powdered cellulose allows to reduce the energy consumption for the movement of cellulose-containing raw materials inside the unit with the provision of sealing. Also, the advantage of this unit is that by combining all the process stages, the consumption of reagents is significantly reduced in comparison with the fact that traditional pulping processes would be used as a stage for isolating cellulose from plant material. Due to lower consumption of reagents, harm to the environment is reduced.

Acknowledgments
The submissions were received during the execution of the state task within the framework of the "Initiative scientific projects" event on the topic No. 13.5443.2017 / BC "Modification of the physical and chemical properties of wood biomass influencing the performance characteristics of the created materials".

References

[1] Ching, T. W., Haritos, V., & Tanksale, A. (2017). Microwave assisted conversion of microcrystalline cellulose into value added chemicals using dilute acid catalyst. *Carbohydrate Polymers*, *157*, 1794-1800.

[2] D Diop, C. I. K., & Lavoie, J. M. (2017). Isolation of nanocrystalline cellulose: A technological route for valorizing recycled tetra pak aseptic multilayered food packaging wastes. *Waste and Biomass Valorization*, *8*(1), 41-56.

[3] Song, Y. K., Chew, I. M. L., Choong, T. S. Y., & Tan, K. W. (2016). NanoCrystalline Cellulose, an environmental friendly nanoparticle for pharmaceutical application–A quick study. In *MATEC Web of Conferences* (Vol. 60, p. 01006). EDP Sciences.

[4] Hideno, A. (2016). Comparison of the thermal degradation properties of crystalline and amorphous cellulose, as well as treated lignocellulosic biomass. *BioResources*, *11*(3), 6309-6319.

[5] Tianchun, L. X. X. W. L., & Motoyuki, S. A. S. (2001). Study on the Continuous Transportation of Biomass under High Pressure.

[6] Auxenfans, T., Crônier, D., Chabbert, B., & Paës, G. (2017). Understanding the structural and chemical changes of plant biomass following steam explosion pretreatment. *Biotechnology for Biofuels*, *10*(1), 36.

[7] Pielhop, T., Amgarten, J., Rohr, P. R., & Studer, M. H. (2016). Steam explosion pretreatment of softwood: the effect of the explosive decompression on enzymatic digestibility. *Biotechnology for Biofuels*, *9*(1), 152.

[8] Sui, W., & Chen, H. (2016). Effects of water states on steam explosion of lignocellulosic biomass. *Bioresource Technology*, *199*, 155-163.

[9] Prosvirnikov, D. B., Safin, R. G., Ziatdinova, D. F., Timerbaev, N. F., & Lashkov, V. A. (2016, April). Multifactorial modelling of high-temperature treatment of timber in the saturated water steam medium. In *IOP Conference Series: Materials Science and Engineering* (Vol. 124, No. 1, p. 012088). IOP Publishing.

[10] Prosvirnikov, D. B., Ziatdinova, D. F., Timerbaev, N. F., Saldaev, V. A., & Gilfanov, K. H. (2016, April). Mathematical modelling of the steam explosion treatment process for pre-impregnated lignocellulosic material. In *IOP Conference Series: Materials Science and Engineering* (Vol. 124, No. 1, p. 012087). IOP Publishing.

[11] Sadrtdinov, A. R., Sattarova, Z. G., Prosvirnikov, D. B., & Tuntsev, D. V. (2015, December). Modeling of thermal treatment of wood waste in the gasifiers. In *Mechanical Engineering, Automation and Control Systems (MEACS), 2015 International Conference on* (pp. 1-5). IEEE.

[12] Oliveira, F. M., Pinheiro, I. O., Souto-Maior, A. M., Martin, C., Gonçalves, A. R., & Rocha, G. J. (2013). Industrial-scale steam explosion pretreatment of sugarcane straw for enzymatic hydrolysis of cellulose for production of second generation ethanol and value-added products. *Bioresource technology*, *130*, 168-173.

[13] Rocha, G. J., Martín, C., da Silva, V. F., Gómez, E. O., & Gonçalves, A. R. (2012). Mass balance of pilot-scale pretreatment of sugarcane bagasse by steam explosion followed by alkaline delignification. *Bioresource Technology*, *111*, 447-452.

Use of geonics scientific positions for designing of building composites for protective (fortification) structures

R S Fediuk[1,a], Yu G Yevdokimova[1], A K Smoliakov [1], N Yu Stoyushko [1], V S Lesovik [2]

[1]Far Eastern Federal University, Vladivostok, Russia
[2]Belgorod State Technological University named after V.G. Shukhov, Belgorod, Russia

E-mail: [a] roman44@yandex.ru

Abstract. The examples of the implementation of the geological (geomimetic) positions in construction materials science are given in the work. The wall materials obtained with this technology have a much more developed surface than traditional wall materials. The second example of using such approaches is the development of internal care systems that will create a more highly organized structure of cement stone at the macro-, micro- and nano-scale levels and concrete in general at all stages: the stage of mixing, hardening and exploitation of the material. The regularities of the structure formation processes are revealed and the principles for increasing the efficiency of non-autoclaved aerated concrete due to application of composite binders, process control in a three-phase disperse porous system and the development of technological methods for the production of protective (fortification) structures are developed.

1. Introduction

A feature of modern science, is a narrow specialization. Of course, differentiation contributes to a significant increase in the accuracy and depth of knowledge about a narrow range of phenomena and processes, but at the same time leads to a weakening of the links between individual scientific disciplines. It goes to the point that experts in narrow areas of the same science often do not understand either theories or the methods of investigating each other.

This applies to scientists involved in the problems of the military-industrial complex. In addition, the situation here is complicated by the fact that military developments, although they are the most advanced in comparison with general civilian design, have been classified in the overwhelming majority of cases, so often researchers from different organizations work on one problem, not being able to compare the results with each other And with earlier developments (including embedded ones) [1-3].

At the same time, a number of organizations of the Russian Federation tend to unite scientific potentials and existing developments for the common good - enhancing the country's defense capability. In particular, this refers to the scientific cooperation of the Far Eastern Federal University (Vladivostok) and the Belgorod State Technological University named after V.G. Shukhov (Belgorod) in the field of creation of protective (fortification) structures of a new generation.

In particular, prof. V.S. Lesovik has developed the positions of geonics (geomimetics) - a transdisciplinary science that solves engineering problems taking into account the knowledge obtained in the study of geological and cosmochemical processes and the results of their activities: minerals and rocks. Collaborating to prof. V.S. Lesovik the scientific school, which includes the authors of this article, is developing a new generation of protective (fortification) structures.

2. Results and Discussionon

The movement of association began in science in the industrial society from the late 30-40-ies of the 20 century. Interdisciplinary research began. New knowledge has appeared at the junction of scientific research due to the synergistic effect. This is how geophysics, geochemistry, biotechnology, and new directions in cybernetics-bionics and geonics - emerged. Interdisciplinarity is characteristic for a post-industrial society, both in science and in education. But at the present time we are already talking about transdisciplinarity in the study. In fact, we are currently investigating complex systems. The classical approaches that existed earlier cannot allow new breakthrough results [4-5].

Transdisciplinarity research is based on the large-scale use and transfer of knowledge, patterns, cognitive schemes from one discipline to another, with the emergence of emergent properties, that is, properties that are not possessed by separate links in the system or discipline, but they are a consequence of the effect of system integrity. Cybernetics is one of the first transdisciplinarity areas. Bionics and geonics are the most important areas of cybernetics.

Bionics is a science borderline between biology and technology, solving engineering problems on the basis of analysis of the structure and life activity of organisms (Fig. 1). The geonics, one of the directions of cybernetics appeared at the end of the 20th century. In particular, the geonics can use the results of a study of geological processes for the development of technologies for the production of materials.

Fig. 1. Bionics - the application in the technology of biological structures, the most perfect in fundamental relationships and the most cost-effective "material"

Mankind is currently on the threshold of a very difficult stage of its development. The beginning of the 21st century is characterized by such problems as depletion of hydrocarbon reserves, environmental degradation, lack of fresh water, activation of natural and man-made disasters. Over the past 20 years, the number of natural processes disastrous in consequences has increased 4-fold, the number of floods 6-fold, the number of people affected by emergencies has increased from 2 to 154 million a year. Accordingly, it is necessary to develop protective structures of various types that can provide comfortable conditions for human life.

We copy geological processes, creating building materials. For example, an analog of fine-grained concrete is a sandstone (Fig. 2*a*), an analogue of concrete on a large aggregate is a conglomerate (Fig. 2*b*), and so on. But creating building materials we cannot yet achieve the quality and properties of natural analogues. For example, foamed glass with a density of 250-300 kg/m^3 has a compressive strength of at most 1.5 MPa, and porous basalts, volcanic porous rocks with the same density have a strength of 5-6 times greater. The same can be said for cellular concrete of hydration hardening or, for example, masonry solutions - this is an anisotropic texture that completely replicates (resembles) banded rocks. But if the mortar mortars in the structures have a tensile strength of about 1 MPa, the banded rocks have a tensile strength of up to 70 MPa. It is the study of the microstructure of rocks that has enabled us to significantly improve the quality of similar materials. Based on the imitation of natural processes, we created a wall material of increased strength, gas tightness and water resistance, we developed a wide range of composites using these rocks.

a) *b)*

Fig. 2. Natural sandstone (*a*) and natural conglomerate (*b*)

Intelligent building composites are offered - these are composites in which a system of interaction with the environment that positively influences the system "man-material-environment" is laid, these are composites of the future.

Prospectivity of the geological approach in the creation of intelligent building materials can be confirmed by such properties as:

- increased protective and physicomechanical characteristics of composites;
- adaptability of properties, structures to living conditions (different construction zones);
- energy saving, reduction of negative impact on the environment;
- multifunctionality of constructive solutions;
- a wide opportunity for regeneration (self-cleaning of the surface, "healing" of cracks, etc.).

Intellectual materials are composites that, when designed, contain a system of interaction with the environment that allows materials to react to external influences and positively influence the triad "man-material-environment". An algorithm for controlling the processes of structure formation in the development of intelligent composites is developed. When designing intelligent building composites, it is necessary to control the processes of structure formation at all levels, which will allow the material to react to emergence of extreme situations when operating protective (fortification) structures.

The design of the given structure of the material at the nanoscale is carried out with the help of the quadaron approach, and also with the method of crystal seeding; at the micro level due to the creation of crystallization centers and the regulation of properties by additives introduced; at the macro level, rational selection of components, pore filling and empty, and by creating the closest packages is necessary (Fig. 3).

Fig. 3. The designing of the material at nano level

Nanopowders with a specific surface 1500 times greater than that of cement were obtained from the hydrothermal sources of Kamchatka. Additition of a nanopowder in an amount of 0.01% of the cement mass made it possible to increase the strength of high-quality concrete by a factor of 2 [6].

The condition of the contact zone between the core and the matrix largely determines the physico-mechanical properties of the composite material. It is known that the character of the course of the processes of structure formation in the contact zone is determined by adhesion bonds, the saturation of the active centers of the surface of the aggregate, and other factors that are of great importance in the formation of silicate products. The possibility of influencing the processes of further formation of the strength properties of the contact zone is very limited and in most cases the cause of the destruction of silicates under load is due to insufficient strength precisely in these areas. The contact zone determines the strength characteristics of silicates with aggregates also when they are used in corrosive environments and in conditions of cyclic freezing and thawing [7-9].

The basicity of the resulting hydrosilicates is easily regulated by the ratio of silica with alkali metal hydroxide in the core of the active granule. Such particles easily penetrate into the silicate matrix surrounding the granule, interact with portlandite, bind it to insoluble slow-moving complex compounds of calcium hydrosilicates. At the same time, a zonal increase in the true density of the silicate matrix is fixed from 2.21 to 2.54 g /cm^3, cracks and micropores are overgrown (Fig. 4).

The method for grinding a crystalline material, for example quartz sand, significantly affects the kinetics and completeness of the dissolution of the granular aggregate obtained therefrom. The use of a ball mill minimizes the crystal structure of the material to a minimum. The use of a planetary mill or a vibrator is more likely to disrupt the crystalline structure in the grinding process, which speeds up the dissolution processes of the granules. In the case of amorphous silica - perlite, flask, rock Tripoli, etc., mechanoactivation does not significantly accelerate the formation of sodium hydrosilicates and depends mainly on the amount of fine particles of amorphous silica.

Building materials science, as well as geology (geomimetics), pay great attention to the creation of new high-strength, smart materials. They should have an ordered structure, formed neoplasms have high strength and low permeability, have the ability to self-heal structural defects and eliminate softening porosity. These properties are possessed by our silicate products at the stage of their manufacture and autoclave processing [10-11].

Fig. 4. Microstructure of cement composite

Another new approach is the use of composite sinter-based binders, which create the most favorable conditions in the early stages of structure formation and hardening of the system. This leads to a reduction in stresses in the hardening composite and, as a consequence, to a decrease in the number and size of microcracks, which determines the technical and economic effectiveness of using a sinter-based composite to produce a concrete mixture used in dry hot climates. It is known that volcanic tuff is heteroporous rock. The pore space of this rock is very complex in its form and consists of a combination of pores of various sizes. Under operating conditions under various loads, microcracks self-liquidate due to the interaction of moisture contained in tuff particles with unreacted clinker minerals. Particles of tufa in the process of exploitation give up their stored capillary-retained water, and this will lead to activation of the processes of structure formation and the synthesis of a denser homogeneous structure of materials during the hardening and exploitation of concretes, these are so-called intelligent composites.

Introduced by prof. V.S. Lesovik, the concept of "technogenic metasomotosis" is a complex mechanism. This is recrystallization, and diffusion, and dehydration, crystal-chemical transformations, etc. And the whole complex of complex physico-chemical transformations, that is, technogenic metasomotosis, when these phenomena are taken into account when designing the material, will allow the material, reacting to external influences, to self-heal defects that are formed during operation and restore its original characteristics.

The next step is development of the intelligent composites of a new generation with a hybrid matrix. In particular, the results obtained in the study of meteorites can be used to create hybrid composites with unique properties.

Taking into account the position of the geonics (geomimetics) prof. V.S. Lesovik proposed "The law of affinity of structures". It implies the design of layered composites and repair mixtures at nano-, micro- and macro- levels similar to the base matrix, which leads to a significant increase in adhesion and durability of materials.

The fundamental basis for designing materials for various purposes is the physico-chemical approach to understanding the aggregate of the structural and associative properties of microscopic formations: compounds, molecules, atoms, but even knowledge of the microstructure is insufficient to predict the final microscopic properties of the system being created. To assess the characteristics of the material, the entire complex of interactions, both its individual components and structural organizations, is taken into account. Considering the essence of the affinity of building composites, we mean the concept and properties of the chemical affinity of elements, Speaking of building materials,

IOP Conf. Series: Materials Science and Engineering **221** (2017) 012011 doi:10.1088/1757-899X/221/1/012011

we first of all take into account the chemical composition of the constituent elements from which these materials are created. But the main criterion for assessing the characteristics of affinity, according to prof. V.S. Lesovik and prof. L.Kh. Zagorodnjuk, is the strength and durability of the contact layer. When creating a material, taking into account the affinity of the structures, it is necessary to take into account the structural features of the structure of the material itself and take this into account when laying it in the structure, as well as in its further operation.

For example, concrete scrap from products and structures that have been in operation for decades, especially after steaming, after grinding does not show sufficiently active binding properties due to the almost complete hydration and carbonization of clinker minerals and hydrate phases. However, even the products of the crushing of the oldest scrap differ by a much higher physicochemical affinity with the cement matrix of the concrete than conventional disperse additives such as quartz or limestone sand, screening of granite and other acidic silicates, and the like. In this regard, ground crushing screening, especially characterized by an increased content of Portland cement component, can be attributed to the active fillers of the binder and concrete mixture as a whole. It is known that when cement is mixed with various technogenic products in the presence of a superplasticizer, which facilitates the process of the binding of the binder and prevents the aggregation of the crushed cement particles, astringents with increased chemical activity and reduced water demand with a specific surface of 450-550 m^2/kg are obtained.

The theoretical basis for improving the physico-mechanical characteristics and durability of building materials can be technogenic metasomotosis in building materials science. This is a stage in the evolution of building materials, characterized by the adaptation of the composite to changing conditions during the operation of protective (fortification) structures. This chemical interaction in the system "Binder-aggregate-filler-additive-pore solution-environment" with a change in the chemical composition, in which dissolution of the original components and the synthesis of tumors occurs almost simultaneously. Technogenic metasomotosis is a complex mechanism. This is recrystallization, and diffusion, and dehydration, crystal-chemical transformations, transformation in the crystal lattices of the mineral, autogenesis (separation of the solid phase and solutions).

Additive technologies or layer-by-layer synthesis technologies are one of the most dynamically developing areas in science today. For a fairly short process of time that has passed since the appearance of the 3D printer, people have learned to print dishes, toys, cars and even human organs and tissues. The range of items that can be printed using a three-dimensional printer is constantly expanding.

Without taking into account the changes taking place in the environment, now it is impossible to pass to the development of 3D additive technologies. The fundamental basis for the design and creation of composites for additive technologies is the transition to transdisciplinarity research, which is now leading, including the direction of geonics (geomimetics). Examples of additive technologies can often be found in nature. For example, masonry solutions are a layer-by-layer synthesis. Brick masonry is an analog of layered rocks. But if the mortar mortars in the structures have a tensile strength of about 1 MPa, the banded rocks have a tensile strength of up to 70 MPa. It is the study of the microstructure of rocks that has enabled us to significantly improve the quality of similar materials.

The introduction of additive technologies (layer-by-layer application) is impossible without the use of the "The law of affinity of structures". It involves the design of layered composites and repair systems at the nano-, micro- and macro-level similar to the base matrix, which leads to a significant increase in the adhesion and durability of materials. When we need to create a single whole composite and thus only the law of affinity of structures allows us to do this.

3. Conclusion

The examples of the implementation of the geological (geomimetic) positions in construction materials science are given in the work. The wall materials obtained with this technology have a much more developed surface than traditional wall materials. The second example of using such approaches is the development of internal care systems that will create a more highly organized structure of cement stone at the macro-, micro- and nano-scale levels and concrete in general at all stages: the stage of mixing, hardening and exploitation of the material. The regularities of the structure formation processes are revealed and the principles for increasing the efficiency of non-autoclaved aerated concrete due to application of composite binders, process control in a three-phase disperse porous system and the development of technological methods for the production of protective (fortification) structures are developed.

Implementation of the concept of designing the building composites of the future made it possible to create high-strength concretes with a compressive strength of up to 200 MPa.

References

[1] Ibragimov R. The influence of binder modification by means of the superplasticizer and mechanical activation on the mechanical properties of the high-density concrete. ZKG International, 2016. No69 (6) pp.34-39.
[2] Zagorodnjuk L.H., Lesovik V.S., Volodchenko A.A., Yerofeyev V.T. Optimization of mixing process for heat-insulating mixtures in a spiral blade mixer. International Journal of Pharmacy and Technology, 2016. No.8(3) pp.15146-15155.
[3] Ibragimov R.A., Pimenov S.I. Influence of mechanochemical activation on the cement hydration features. Source of the Document Magazine of Civil Engineering, 2016. No.62 (2) pp 3-12.
[4] Glagolev E., Suleimanova L., Lesovik V. High reaction activity of nano-size phase of silica composite binder. International Journal of Environmental and Science Education, 2016. No.11(18) pp.12383-12389.
[5] Fediuk R.S. Mechanical Activation of Construction Binder Materials by Various Mills. IOP Conference Series: Materials Science and Engineering, 2016. No.125(1) pp.012019.
[6] Lesovik V.S., Potapov V.V., Alfimova N.I., Ivashova O.V. Increase in the effectiveness of binders due to the use of nanomodifiers. Building materials, 2011. No. 12 pp. 60- 62.
[7] Cai J., Hu X., Xiao B., Zhou Y., Wei W. Recent developments on fractal-based approaches to nanofluids and nanoparticle aggregation. International Journal of Heat and Mass Transfer, 2017. No.105 pp. 623-637.
[8] Morsy M.S., Al-Salloum Y.A., Almusallam T.H., Abbas H. Mechanical properties, phase composition and microstructure of activated Metakaolin-slaked lime binder. KSCE Journal of Civil Engineering, 2017. No.21(3) pp. 863-871.
[9] Logbi A., Kriker A., Snisna Z. Effects of mineral additions on durability and physico-mechanical properties of mortar. AIP Conference Proceedings, 2017. No.1814 pp. 020032.
[10] Parthiban K., Saravana Raja Mohan K. Influence of recycled concrete aggregates on the engineering and durability properties of alkali activated slag concrete. Construction and Building Materials, 2017. No.133 pp. 65-72.
[11] Mohammadinia A., Arulrajah A., Sanjayan J., Bo M.W., Darmawan S. Strength development and microfabric structure of construction and demolition aggregates stabilized with fly ash-based geopolymers. Journal of Materials in Civil Engineering, 2016 No. 28(11) pp.04016141.

IOP Conf. Series: Materials Science and Engineering **221** (2017) 012012 doi:10.1088/1757-899X/221/1/012012

Combined Delivery of Consolidating Pulps to the Remote Sites of Deposits

V I Golik[1, 2, a], **A B Efremenkov**[3, b]

[1]North-Caucasian State University of Technology, 44 Nikolaev street, Vladikavkaz, 362021, Russia[1]

[2]Vladikavkaz Scientific Center of the Russian Academy of Sciences and the Government Republic of North Ossetia-Alania, 22, Markusa street, Vladikavkaz, 362027, Russia

[3]Yurga Institute of Technology, TPU Affiliate, 26, Leningradskaya street, Yurga, 652055, Russia

E-mail: [a] v.i.golik@mail.ru, [b] abe73@rambler.ru

Abstract. The problems of modern mining production include limitation of the scope of application of environmental and resource-saving technologies with application of consolidating pulps when developing the sites of the ore field remote from the stowing complexes which leads to the significant reduction of the performance indicators of underground mining of metallic ores. Experimental approach to the problem solution is characterized by the proof of technological capability and efficiency of the combined vibration-pneumatic-gravity-flowing method of pulps delivery at the distance exceeding the capacity of current delivery methods as it studies the vibration phenomenon in industrial special structure pipeline. The results of the full-scale experiment confirm the theoretical calculations of the capability of consolidating stowing delivery of common composition at the distance exceeding the capacity of usual pneumatic-gravity-flowing delivery method due to reduction of the friction-induced resistance of the consolidating stowing to the movement along the pipeline. The parameters of the interaction of the consolidating stowing components improve in the process of its delivery via the pipeline resulting in the stowing strength increase, completeness of subsurface use improves, the land is saved for agricultural application and the environmental stress is relieved.

1. Introduction

The industrial demand for the mineral raw materials is growing due to the population growth and the changes in geography of mining. It can be satisfied not only with the use of new technical facilities and modern technologies but also due to application of nonborrowed reserves of production. One of the trends of mining modernization is simplification of the methods of stowing pulps production and delivery to the place of its application [1-2].

During the underground extraction of the vast majority of metal ore deposits the scope of application of technologies using the consolidating stowing for controlling the state of the ore-hosting

IOP Conf. Series: Materials Science and Engineering **221** (2017) 012012 doi:10.1088/1757-899X/221/1/012012

massifs is limited to developing parts of the ore field remote from the stowing complexes, that is why implementation of operating stowing complexes to satisfy the demand for the stowing materials is becoming a more and more relevant problem.

2. Materials and methods of research

The traditional methods of delivering the consolidating stowing from the stowing complex to the place of application through the pipes are as follows: gravity-flowing, mechanic, pneumatic and combined including the elements of two-three methods.

Transportation of the stowing materials through the pipes according to the gravity-flowing method depends upon the properties of the material and upon the horizontal and vertical components of the pipeline. The distance of consolidating stowing delivery often achieves 1500 meters under the favorable grade.

The delivery distance increases if gravity-flowing is completed with supply of compressed air into the pipeline. In the pipeline volumes of the consolidating stowing pulp are formed divided by air gaps and the speed of the material movement increases up to 25 m/s.

Both methods are reliable when the relation of the vertical and the horizontal parts does not exceed 1/5.

When developing the deposits remote from the main site it is necessary to deliver the materials at the distance over 1500 m under the small height of the vertical part of the pipeline [3].

Such problem turned up when developing the closely situated North Kazakhstan deposits Shokpak and Kamyshovoye. In the neighborhood of the deposits fertile land is situated making any methods of deposit management impossible beside application of consolidating stowing. At the same time building of the second stowing complex rose the price of ore extraction making it economically unacceptable [4].

To prove the technological capability and economic feasibility of the new technology includes modeling of the processes which confidentiality of results increases when applying the materials of on-the-farm research under the industrial conditions.

The practice of stowing materials delivery with increase of the gravity-flowing capability due to its combining with other methods is relatively small. For example, at the deposit "Vismut" (Germany) the consolidating stowing material was supplied by vibro-gravity-flowing method at the horizontal distance which three-fold exceeds the height of the vertical flight [5-6].

Vibration-pneumatic-gravity-flowing method differs fundamentally from the known methods due to the application of the phenomenon of the pipeline vibration on elastic support with the help of vibration generators ensuring its vibration.

3. Experiment

Installation of vibration-pneumatic-gravity-flowing transport which was first introduced in the USSR at the deposit Shokpak-Kamyshovoye (North Kazakhstan) united the horizontal and the vertical parts of the pipeline (Fig.) [7]. The horizontal part of the pipeline was set at rubber support. Each section of the pipeline 200 meters length was supplied with inertia single-shaft vibration generator with electric drive and pneumatic tie-in with electromagnetic valve. The sections of the horizontal pipeline were connected with elastic connecting parts to exclude influence of one section upon another.

The volume of separate portions of the consolidating stowing was changing within the limits of 200…400 m^3.

After filling of the stage site and the first section of the horizontal site with the stowing material the vibration generator was turned on beginning with the first one which reduced the resistance of the material to the transportation.

The measured difference in pressure through the length of the section made: under gravity-flowing 0.6-1.0 MPa, grad P =3.0-5.0 kPa/M, and under the vibration mode – 0.12-0.20 MPa, grad P = 0.8-1.0 kPa/M.

IOP Conf. Series: Materials Science and Engineering **221** (2017) 012012 doi:10.1088/1757-899X/221/1/012012

When the stowing material was moving at a speed of 1.0-1.5 m/s in the pipeline the components of the pulp were redistributed with decomposition of the granules and prevention of lamination which improved its strength by 20-25%.

Energy consumption of the stowing material when transporting it at the distance of up to 2.5 km, vibration frequency of 10-30 Hz, amplitude 0.5-1.5 mm made 0.15-0.22 kW/h for 1 m^3.

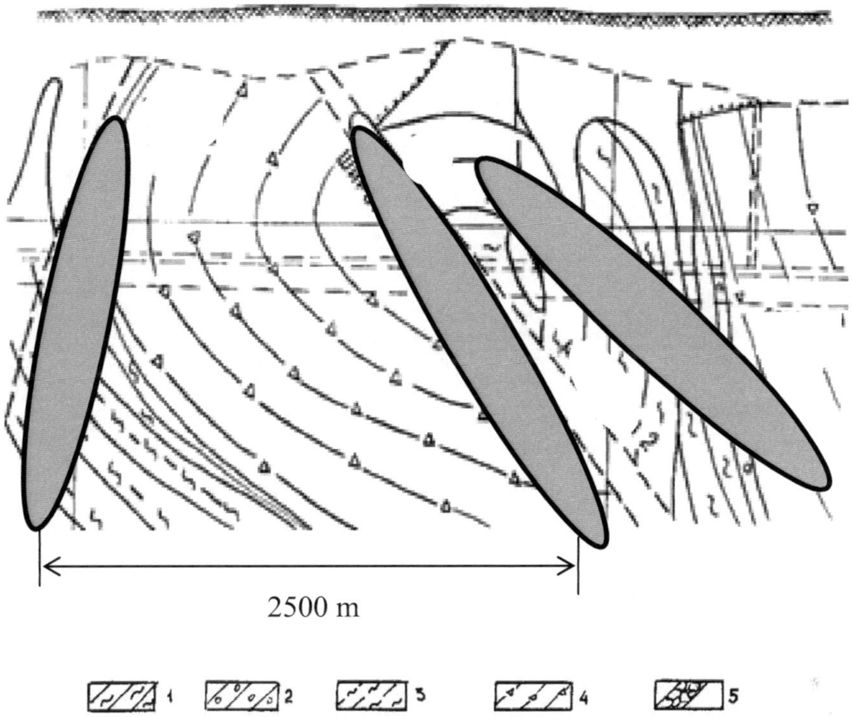

Fig.1. Deposit Shokpak-Kamyshovoye: 1-jasper-quartz rock; 2-conglomerates; 3-siltstone; 4-gravelstone, breccia; 5-zone of rock crush

The properties of materials for producing consolidating stowing: sand from the local open-cut: density 2.64 t/m^3; content of dust-like, clayey and silty 10 30 %; 5.0 mm sieve residue is 1%; specific surface – 6.9 m^3/kg without taking the levigated particles into consideration.

For gravity-flowing to increase the transportability water is added into the pulp until cone slump 11 which increases the consumption of binding materials and the cost of the pulps. And with the help of the vibration generator the stowing was transported even under the cone slump 9.

The differential pressure in the stowing line 1200 m long:

- productivity 80 m^3/h:12.03 – 4.55 = 7.48 atm.;
- productivity 100 m^3/h:15.0 – 5.38 = 9.62 atm.

Specific resistance (ΔP_C):

- productivity 80 m3/h: $$\Delta P_C = \frac{7.48}{1200} = 0.00623$$ atm. or 62kg/m2/1 m; (1)

- productivity 100 м3/ч: $$\Delta P_C = \frac{9.62}{1200} = 0.008$$ atm. or 80 kg/m2/1 m. (2)

- for productivity 80 m3/h and 100 m3/h, accordingly:

$$\tau 0 = \frac{62.3 \cdot 0.17}{4} = 2.6\ \text{kg/m}^2 \text{ and } \tau 0 = \frac{80 \cdot 0.17}{4} = 3.4\ \text{kg/m}^2 \quad (3)$$

To monitor the pressure in the stowing pipeline manometers were installed along the pipeline route in the pickets from RM1 to RM9 (Fig. 2). Air cut-ins were installed near the vibration generators. Under the reverse slope of the pipeline 0.003 the stowing pulps are transported by gravity-flowing until picket 7 (1580 m) and then – with application of vibration generators.

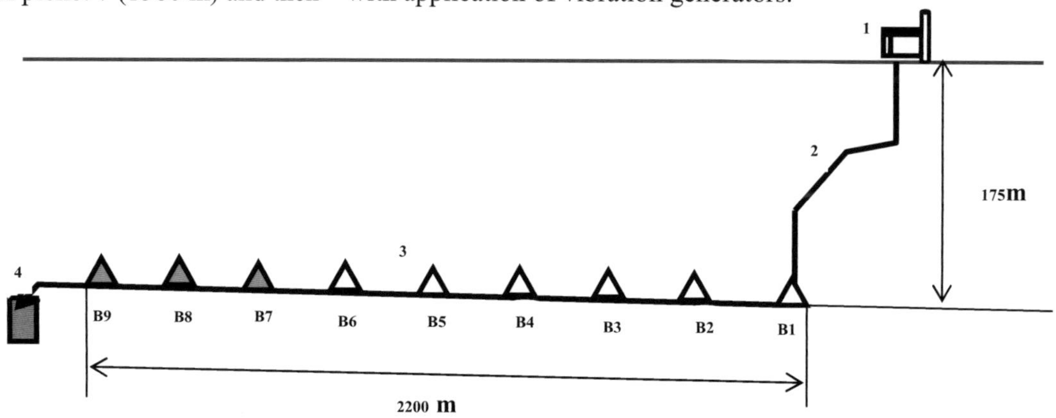

Fig.2. The diagram of vibration-pneumatic-gravity-flowing transportation of consolidating pulps from the stowing complex of Shokpak deposit to the shrink stoping of the deposit Kamyshovoye: 1 – stowing complex; 2 – vertical part of the pipeline; 3 – vibration generators; 4 – the stowing chamber of the block; B1-B9 – single-shaft vibration generators with electric motor drive

The limit length of the gravity-flowing transport is 1600 m; the experimentally found length of vibration-pneumatic transport is 600 m; the delivery distance of the pulps is 2200m.

During the experiment set up the pressure in the stowing pipeline was measured in points 1, 7, 8 and 9 simultaneously at three minute interval separately under the productivity of the stowing plant of 80 and 100 m^3/h.

At first the pulps were transported with application of air without the vibration generators. Up to picket 7 at the distance of 1500 m the stowing materials were successfully transported by gravity-flowing. From picket 7 to the place of the deposit the stowing pulps were transported with application of vibration generators to the distance of 600 m in horizontal direction.

Air pressure in the air and stowing lines:

- under the productivity of 80 m^3/h:

P_M= 60800 kg/m^2 (6.08 atm.); P_T= 45500 kg/m^2 (4.55 atm.);

- under the productivity of 100 m^3/h:

P_M = 67400 kg/m^2 (6.74 atm.); P_T = 53800 kg/m^2 (5.38 atm.).

The speed of the stowing material movement:

- for the productivity of 80 m^3/h: 6.7 m/s;
- for the productivity of 100 m^3/h: 5.7 m/s

The length of the portions of the stowing pulp:

- for the productivity of 80 m^3/h:88 m
- for the productivity of 100 m^3/h:128 m

Specific resistance of the stowing to the movement:

- for the productivity of 80 m^3/h:517 kg/m^2 for 1 m;

IOP Conf. Series: Materials Science and Engineering **221** (2017) 012012 doi:10.1088/1757-899X/221/1/012012

- for the productivity of 100 m^3/h:420 kg/m^2 for 1 m.

Correction factors for the speed:
- for the productivity of 80 m^3/h:0.027 s^2/m^2;
- for the productivity of 100 m^3/h:0.24 s^2/m^2.

The increasing of the strength of the stowing after pneumatic-gravity-flowing to the distance of 1500 m is about 11% at 28 days and 13-14 % at 90 days [8].

4. Results and their discussion

The stable operation of the transportation is ensured by the compliance of velocities at the gravity-flowing and pneumatic sites or the equality of the time of the total portion transportation and its filling into the pipeline. The transportation conditions can be regulated either by the plasticity of the pulps or by changing the productivity of the stowing plant. Turning the vibration generators up to the possibility frontiers of the gravity-flowing transport increases resistance to the pulp motion, that is why when transporting along the pipeline with reverse slope the vibration generators do not always improve the indicators of consolidating stowing transportation.

For the objective comparison of the opportunities of transport diagrams we compare the results of the mathematical modeling of pneumatic-gravity –flowing and pneumatic-vibration-gravity-flowing transportation under similar productivity of the complex (80 m^3 /h) and delivery distance of 1500 m which is critical for the first scheme and starting for the second one. Under the pneumatic-gravity-flowing transportation to the distance of up to 1500 m strength increasing of the stowing pulp reaches 3% at 28 days and 6% at 90 days and under pneumatic-gravity-flowing transportation to the same distance it reaches 11% at 28days and 14% at 90 days.

All transportation conditions being equal the new technology reduces the resistance to the motion, increases the length of the delivered portion of the pulp and improves the strength of the consolidating stowing due to the vibration of the pipeline walls. The efficiency of the consolidating stowing activation under the alternative layouts is characterized by the values:
- for the pneumatic-gravity-flowing delivery – 1.03-1.05;
- for the vibration-pneumatic-gravity-flowing delivery – 1.05-1.20.

In the case of pulps transportation to Kamyshovoye deposit from the operating stowing complex at Shokpak deposit there is no necessity to build a new stowing complex with corresponding capital costs which are replaced by much smaller costs associated with the pipeline construction. As a result of the pulp activation the cement consumption can be reduced leading to reduction of the pulp cost price by 15%.

When transporting the consolidating stowing according to the new technology the completeness of the use of subsurface resources increases and the ecological stress is reduced.

The efficiency of the consolidating stowing transportation technology is estimated through comparison of the reduced costs [9-12]:

$$E = [(C_1 + E_h K_1) - (C_2 + E_h K_2)] \cdot A_y \tag{4}$$

where E – annual cost advantages, rub; E_h – ratio of discounting expenditures and profit in the course of time, unit fractions; K_1 and K_2 – capital investments into production funds under building a new stowing complex and combined delivery accordingly, rub; C_1 and C_2 – producing cost of 1 t of concentrate under the compared variants accordingly, rub; A_y – annual production after involving new deposit resources into operation. Combining of the methods of pulps delivery is an element of the mining enterprises economy management system under the intensification of competition at the market and surviving without the governmental support [13-15].

5. Conclusions

1. The scope of application of the development systems implementing stowing increases if the consolidating stowing is supplied to remote sites of the deposit according to the combined diagram with application of vibratory technology.

2. Application of vibration generators improves the opportunities of pulp transportation to the distances exceeding the capabilities of traditional pneumatic-gravity-flowing delivery due to reduction of friction on the walls of the pipeline.
3. The consolidating stowing activated in the process of delivery via the pipeline is characterized by a more uniform distribution of the filler which results in its better strength.

References

[1] Magomedov Sh.Sh. Preparing and transporting the consolidating pulps. / Collected works: Math. Int. Conf. "Logical management of technical processes and systems" Moscow Vladikavkaz. 1999. P.65-73.
[2] Cavalcante P.R.B., Palkovits F. Pastefill — asafetysolutionforpillarmining / Australian Centrefor Geomechanics. Perth, 2011. 443 p.
[3] Gridley N. C., Salcedo L. Cemented paste production provides opportunity for underground ore recovery while solving tailings disposal needs / Australian Centre for Geomechanics. Perth, 2011. 431 p.
[4] Golik V.I. SPECIAL METHODS OF DEPOSIT DEVELOPING // Moscow, 2014. Ser. Higher education: Baccalaureate, 132 p.
[5] Platonov V.N., Podubny I.K. Construction and operation of vibration-gravity-flowing of the plant intended to be used for delivery of the stowing at the ore mine in Thuringia. Vibration engineering. / Seminar materials. – M.: Obshchestvo "Znaniye". 1992.
[6] Golik V.I. The innovative diagram of bulk cargo loading onto a sea-craft. Sea transport operation. Novorossiysk. 2014. 1.C.56-62.
[7] Lyashenko V.I. Environmental technologies of operating complicated structure mineral deposits// Mining Surveyors newsletter. 2015. – 1. –C. 10–15.
[8] Kaplunov D.R., Rylnikova M.V., Radchenko D.N. Developing the resources portfolio of mining enterprises on the base of the integrated use of the mineral resources of the deposits. Gorny zhurnal. 2013. 12. P. 29-33.
[9] Golik V.I., Polukhin O.N., Petin A.N., Komashenko V.I. THE ECOLOGICAL PROBLEMS OF ORE DEPOSITS DEVELOPMENT OF KMA// Gorny zhurnal. 2013. 4. C. 91-94.
[10] Shestakov V.A., Razorenov Yu.I., Gabarev O.Z. Production quality management at mining enterprises // Recommended by the educational and methodological association for mining education of the Ministry of Education of the RF as a study guide for the students of higher educational institutions training for the specialty 090200 / Novocherkassk, 2001, 262 p.
[11] Vorobiev A.E., Razorenov Yu.I., Ignatov V.N., Dzhimieva R.B. Innovative geotechnologies of bituminous shale and highly viscous oil deposits development // study guide for magisters studying for mining-and-geological and oil specialties / Novocherkassk, 2008, 214 p.
[12] Hasan A., Suazo G., Fourie A. B. Full scale experiments on the effectiveness of a drainage system for cemented paste backfill / Australian Centre for Geomechanics. Perth, 2011. 379 p.
[13] Zarema M. Khasheva and Vladimir I. Golik.The Ways of Recovery in Economy of the Depressed Mining Enterprises of the Russian Caucasus.International Business Management. 2015. 9 (6): pp. 1209-1216.
[14] Adibi N., Ataee-pour M., Rahmanpour M. Integration of sustainable development concepts in open pit mine design // J. Clean. Prod. 2015. Vol. 108.PartA. P. 1037–1049.
[15] Semidotsky V.A., Khasheva Z.M., Gurfel L.I. The trends of governmental regulation of efficient realization of mega-projects (on the example of Sochi-2014). In the collection of works: Economic and law aspects of realization of Russia modernization strategy: real imperatives of dynamic social and economic development. International research-to-practice conference. Ed. by: G.B. Kleiner, E.V. Sobolev, V.V. Sorokozherdiyev, Z.M. Khashcheva. 2014. P. 290-292.

IOP Conf. Series: Materials Science and Engineering **221** (2017) 012013 doi:10.1088/1757-899X/221/1/012013

Control of Rock Mechanics in Underground Ore Mining

V I Golik[1,2, a], A B Efremenkov[3, b]

[1]North-Caucasian State University of Technology, 44 Nikolaev street, Vladikavkaz, 362021, Russia[1]

[2]Vladikavkaz Scientific Center of the Russian Academy of Sciences and the Government Republic of North Ossetia-Alania, 22, Markusa street, Vladikavkaz, 362027, Russia

[3]Yurga Institute of Technology, TPU Affiliate, 26, Leningradskaya street, Yurga, 652055, Russia

E-mail: [a] v.i.golik@mail.ru, [b] abe73@rambler.ru

Abstract. Performance indicators in underground mining of thick iron fields can be insufficient since geo-mechanic specifics of ore-hosting fields might be considered inadequately, as a consequence, critical deformations and even earth's surface destruction are possible, lowering the indicators of full subsurface use, this way. The reason for it is the available approach to estimating the performance of mining according to ore excavation costs, without assessing losses of valuable components and damage to the environment. The experimental approach to the problem is based on a combination of methods to justify technical capability and performance of mining technology improvement with regard to geo-mechanical factors. The main idea of decisions to be taken is turning geo-materials into the condition of triaxial compression via developing the support constructions of blocked up structural rock block. The study was carried out according to an integrated approach based on the analysis of concepts, field observations, and simulation with the photo-elastic materials in conditions of North Caucasus deposits. A database containing information on the deposit can be developed with the help of industrial experiments and performance indicators of the field can be also improved using the ability of ore-hosting fields to develop support constructions, keeping the geo-mechanical stability of the system at lower cost, avoiding ore contamination at the processing stage. The proposed model is a specific one because an adjustment coefficient of natural and anthropogenic stresses is used and can be adopted for local conditions. The relation of natural to anthropogenic factors can make more precise the standards of developed, prepared and ready to excavation ore reserves relying on computational methods. It is possible to minimize critical stresses and corresponding deformations due to dividing the ore field into sectors safe from the standpoint of geo-mechanics, and using less cost-demanding ways of keeping rock massif stable.

1. Introduction

Ore-hosting deposits, in most cases are rocks and half-rocks with higher acoustic impedance, localized in the developed residuum. Thickness of the rock area with their strength falling around small workings is about first meters, and it far more on the chamber contacts. Inside these areas there is a border zone of weakness.

IOP Conf. Series: Materials Science and Engineering **221** (2017) 012013 doi:10.1088/1757-899X/221/1/012013

Excavation of mineral resources leads to the development of technology-related cavities, which are the reasons for arising stresses and deformations in ore-hosting areas, and can cause even earth's crust destruction over them. In underground mining of thick ore fields the performance indexes are insufficient because of inadequate consideration of geo-mechanic specifics of ore-hosting fields, as a consequence, critical deformations and even earth's surface destruction are possible, lowering the indicators of full subsurface use this way [1]. The reason for it is the available approach to estimating the performance of mining according to ore excavation costs, without assessing losses and damage to the environment when nature destructing technologies are of top importance [2].

The integrity of the earth's surface can be provided by means of residual supporting capacity of fractured rocks. At present, the concepts on strengthening mechanism of rocks can't provide full understanding the conditions how to be involved in operation of the ore-hosting field, limiting, this way, the sphere of applying nature-saving underground mining technologies. Therefore, establishment of correlation between the value of mountain pressure and residual supporting capacity of rocks, as well as development of methods to control rock masses when ore extracting and cavities filling, which are aimed at obtaining sufficient indexes of using the Earth's interior are scientifically and practically important issues.

The approaches to solution of geo-mechanical problems are distinguished by combination of methods for justifying technological capacity and efficiency of improving the technology on the base of geo-mechanical factors. The main idea of decisions to be taken is ability of geo-materials to develop the support constructions of blocked up structural rock block. To provide correctness of outcomes the study was carried out according to an integrated approach involving the analysis of concepts, field observations, and simulation with photo-elastic materials in conditions of particular deposits.

2. Materials and Methods

The theory by Protodyakov M.M. (1933) is the most appropriate one for control of ore-hosting field state, in terms of this theory a working is subject to rock weights only, which are within the natural self-supporting arch at the height far smaller than the mining depth.

This theory was further developed in papers of Russian and foreign geo-mechanics. V.D. Slesarev (1948) found out a parameter of stability, i.e. resistance to failure of rocks forming a truss. A.A. Borisov (1964) joined the theory of arch with stability of a rock layer in the door head. S.V. Vetrov (1975) determined a stable position of working as an equality of blocked up rock strength, forming a hinged arch with a weight not exceeding that of natural self-supporting arch [3].

Maximal allowable outcropping spaces in ore-hosting field:

$$L_F \le L_P; h \le H; k \ge 1$$

$$L_P = f(R_1, d_{1.2}, k_2, в, y) \quad (1)$$

where L_F - maximal actual outcropping space of a cavity roof, m; L_P - limiting arch span, m; h – height of mining affected zone, m; H – depth of mining measured from cavity edges to the bedrock level, m; k – coefficient of safety margin, units; R_1 – rock strength (man-made massif), MPa; $d_{1,2}$ – horizontal and vertical sizes of structural blocks, m; k_2 – safety factor, units; $в$ – rock thickness providing no reliable roof constructions, m; γ - rock density, t/m^3.

The stability of ore-hosting field is ensured in conditions of sufficient mechanical strength of the bottom row of blocked up structural blocks loaded by rock weight within the natural self-supporting arch (Fig. 1).

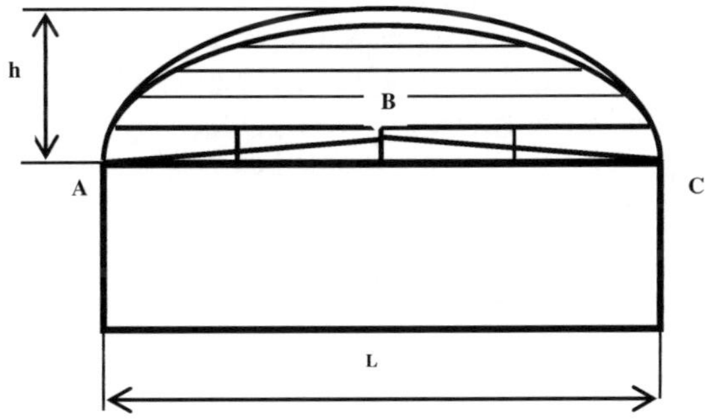

Fig. 1. Operational scheme of three-hinged arch of blocked up rocks: A, B, C – hinges: L – limiting space of level outcropping in the door head; h – height of the natural self-supporting arch

The roof keeps the level form if stresses in the supporting layer do not exceed the ultimate strength of rocks in the contact zone of two rock blocks (Fig. 2).

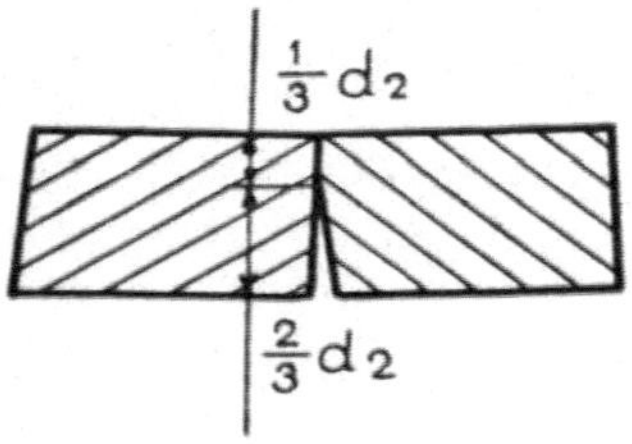

Fig. 2. The condition of a mine working level roof: d_2 – vertical size of the sructuraal rock block

Preservation of the Earth's surface from destruction over the developed mining workings of deposits is provided via controlling the level of natural and technology-related stresses by combining technological processes in time and space. New nature and resource saving technologies and technical means were developed on this base and positive outcomes of their application in underground mining were registered in Russia, Kazakhstan and Ukraine.

Nature and resource saving requirements at mining enterprises consist mainly in the use of low quality mineral resources and wastes in production. Their insufficient activity is improved through recycling. Therefore, optimization of the mining technology on the base of geo-mechanical factors, conversion of geo-materials into the triaxial compression state is an important economic and social problem.

Efficiency of a mining technology is determined via comparing some mining variants, taking into account costs at all processing stages and protection of inhabitants living in the mining-affected zone according to the criterion of Earth's surface safety.

3. Experiment

A promising direction in mining is chamber mining with filling of the mined-out space with hardening stowing mixtures of improved composition and strength.

If the stages in preparing recourses are combined rationally in time and space, a mining company works uninterruptedly and efficiently. The condition of recourses prepared for excavation over those ready for excavation depends on conditions and routine of development, cutting, acquisition of ore recourses in the exploitation block, way of keeping and depreciation of the mined out space.

The blocks are prepared gradually for excavating, that slows down the rates of mining because production capacity of a mining company depends on the volume of recourses ready for excavation.

The division of the ore field into safe zones according to geo-mechanical principles includes into the volume of ready for excavation recourses zones of the field protected from critical deformations by natural constructions of blocked up structural blocks. A group of blocks within such a zone makes it possible to prepare recourses for excavation independently, improving intensity, this way.

Safety of ready for excavation recourses is determined by the possibility of building a construction of blocked up rocks, which keep a flat door head:

$$2a = 2d_1\left(\frac{10R_2}{k_1 yH} - 1\right) \tag{2}$$

where d_l- horizontal dimension of a structural rock block, m; R_2 – ultimate strength of rocks, kg/cm^2; H – depth of mining, m; γ - volume weight of rocks, t/m^3; k_1 – safety margin.

At d_l = 3 m, R_2 = 1400 kg/cm^2; H = 500 m, γ = 2.7 t/m^3 and k_1 = 1, the ultimate width of arch is 54 m.

The equivalent width (L_f) of an ore body with strike *A=500* and dip *B=640 m* is (dimensions are in the plan):

$$L_f = \frac{AB}{\sqrt{A^2 + B^2}} = \frac{500x640}{\sqrt{500^2 + 640^2}} = 394m. \tag{3}$$

The equivalent width exceeds the maximum permissible one; therefore, the ore field is divided into zones 54×54 m via advancing of rooms and filling them with hardening mixtures, the strength of which is calculated according to surcharging with rock mass to the surface level. For example, there are 4 blocks on the area 54×54 m with rated recourses ready for excavation via chamber mining with hardening filling for 12 months. The norm of ready for excavation resources should be 6-12 months.

The most difficult way of mining is mining of thick fields, opening onto the earth surface, a typical example is Tyrnyaus tungsten and molybdenum deposit in the zone of metamorphically altered rocks of upper and middle Paleozoic era and Lower Jurassic, cut with dikes, stocks and sheeted bodies different in age and composition. The ore zone is located among igneous rocks: crystalline schists, pegmatites and granite-gneisses. More than 60% of ore reserves are in the Main Skarn.

The ore body is a saddle-shaped basin with wings falling north and south at an angle of 60-70^0 and eastwards immersing hinge at an angle of 55-60^0 under the thickness of biotite hornblendes. The thickness of the deposit in the inflection is 100-120 m, on the sides it is up to 3-5 m (the average one is 50 m).

The strength by M.M. Protodjakonov and density, respectively, for skarn ores are 16-20 and 3 t/m^3, these of rocks of the hanging side (hornfels) - 12-14 and 2.6 t/m^3, the rocks of the recumbent side (marbles) - 8-12 and 2.5 t/m^3. Ore doesn't tend to oxidation, caking and spontaneous combustion. The degree of ore fragmentation is 1.5.

The main reserves are excavated in induced block caving with blasting holes. The reserves of the main skarn are mined in induced block caving with ore breaking by blasting holes in a clamped medium in one-stage order.

The overlapping rocks break down over the mined blocks on the main skarn and a single crater is formed. There are moving rocks in the pitcrater when breaking arch pillars and blocks.

At a shallow depth of mining the crater has steep edges. As the depth of mining increases, breakage of the hanging side in the thick part of the main skarn becomes smooth. The edges of the crater on the hanging side become flat, and the transverse dimensions of terraces and cracks zone increase. At a depth of 350 m below the surface at some distance from the main pitcrater another crater is formed. Craters above the blocks in the thick part of the main skarn are developed at a speed of 20 to 50 mm/s, having the shape of tubular hollows without characteristic displacement lines.

The complex of studies on geo-mechanical phenomena included instrumental observations and simulation of the process using methods of photo-mechanics and mathematical theory of elasticity. The quantitative estimation of the displacement parameters was made as a result of reference-type station data processing. The method of observation included recording crater dimensions on the earth's surface with registering the dynamics of their development and construction of hazardous zones [4].

The shift in the North-West Skarn was slow. The largest registered subsidence was 149 mm. Horizontal deformations ranged from 8-9 to 3.0 × 10-3. After 1981 the process of rock shifting slowed down. In 1989 the ceiling of block 4 was broken horizontally 2537 m, and the movement of rocks intensified. The largest subsidence was 102 mm. Horizontal deformations in the reference interval are 35-36, in the zone of tectonic disturbance E = 8.3.10-3. The break angle was 76^0.

In 1991 the process of rock shifting became even more active and resulted in opening of rupture cracks on the earth's surface. The largest subsidence (600 mm) was recorded by bench 7; its subsidence rate was 15 mm/s. Over one year 13 craters were formed due to mining with frontal discharge. The standing time of rocks is 10 to 128 days. The volume of ore output 10,000 to 100,000 thousand tons.

The craters appear together with subsidence of the earth's surface at a speed of about 20 mm / s. and occurrence of cracks in the form of closed contours. A clear correlation between time of ore excavation and depth of mining operations was not observed. The craters were formed at a depth of mining from 80 to 550 m. The minimum time of crater formation is 5-10 days, the maximum - 3 years.

The criterion of field state is the ratio of output volume V_p and craters V_f variants are possible:

$$\frac{V_p}{V_f} > 1, \frac{V_p}{V_f} \approx 1, \frac{V_p}{V_f} < 1 \tag{4}$$

In the first case loosened rocks are carried from the release ellipsoid area, cavities are accumulated, and field is subject to intense destruction. As ore is released, a mass of overlying loose rocks starts moving, the earth's surface moves smoothly and slowly, and a shallow crater appears (1.5 m).

In the second case the crater depth increases; the earth's surface breaks rapidly, the period of crater formation is reduced. The peculiarity of rocks breaking process over the mined space is its limited area. As the rock mass is mined out, the rocks and the earth's surface descend. The time of caving rarely exceeds 2-3 weeks. The average depth of craters does not exceed 20 m.

In the recumbent side field of the Main Skarn there were no shifts registered. The average displacement speed of the rocks was 5.7 mm / month in 1991 on its northwestern side, whereas a critical speed was 40 mm/month. The maximum height in the northwestern side of Mukulansky open cast was more than 700 m with a total inclination angle of $26\text{-}27^0$. The height of the ledges is 50-55 m with a slide angle of 50^0 and about 70 m with a slide angle of 40^0.

The breaking zone is a loose medium, characterized by lacking adhesion and weakening surfaces of tectonic cracks type. The calculated angle of edge inclination in it can be equal to the natural slope angle, being conditionally equal to the angle of internal rocks friction of $37\text{-}38^0$ with an unlimited height. The rock displacement process being a consequence of natural structure leads to complete destruction of connection between the structural blocks of rocks and subsidence of large sections with slight disruption of the earth's surface. The zone of rupture influence is determined according to the magnitude of the earth's surface critical deformation - horizontal expansion, and also to the slope and curvature of discontinuous structures arising in underground mining.

The blind deposit was mined out via underground chamber-integral system with dry filling. Chambers 95 m high with dimensions of 30x30 m were separated in plan by inter-chamber pillars of the same dimensions. Thus, chambers 6-7 are located under the bottom of the northern side at a depth of 120 m [5].

The hanging side of the blind deposit, where opencast benches are located, is composed of biotite hornfels (structural blocks are 0.5-1.5 s). Structural blocks are formed, as a rule, as three

IOP Conf. Series: Materials Science and Engineering **221** (2017) 012013 doi:10.1088/1757-899X/221/1/012013

systems of cracks with a strike for the first system of 50-60^0, an angle of incidence of 55-60^0; for the second system - 100-110^0 and 60-70^0, and for the third system - 170-180^0, 25-30^0.

In chamber 7 the roof was broken and the crater arose on the surface of the opencast. When ore mining the crater was expanded, reaching a width of 50 m along the strike and 30 m across the strike. The rock sedimentation rate was 2.7 m/s. After crater formation the surface over the chambers sagged by 5.5 m, and on the mountains 2435 m the settling rate gradually decreased to 2 mm/s.

In case of subsidence crater formation in underground mining workings the safety factor is reduced by up to 30%. The obtained analytical dependencies are verified on models using the methods of photo-mechanics and mathematical theory of elasticity [6].

The models were made of optically active polyurethane with the value of the stripe $\sigma_H^{1,0}$ = 75.6 kg/cm^2 for the conditions: $H = 350m$ - depth of the workings from the day surface; γ = 3.0 t/m^3 – density of the overlapping rocks. To obtain the normal stress on the free edge of the working number of the band order is multiplied by the band value ($\sigma_H^{1,0}$).

Strength condition of the edge:

$$\sigma_1 - \sigma_2 \geq \sin\delta.(\sigma_1 + \sigma_2) + \sigma_3.(1 - \sin\delta), \tag{5}$$

where $\sigma_1(\sigma_2)$ – value of stresses in the edge point; δ =30^0 – inner friction angle; σ_3 = 140-160 MPa.

The models were stressed with lateral extension coefficients 0.5; 1.0; 1.5. The values of the lateral extension are specified according to the features of Tyrnyauz deposit located in the activity zone of the Elbrus mountain system, it redistributes lateral pressure when height vs. width of the chamber > 1.

The problem of estimating the stability of an inter-chamber pillar with decreasing width from 30 to 20 m was solved by modeling. Variants of the chamber section shape were examined to improve or reduce stresses along the edges and in the locks without changing roof dimensions and shape and keeping the same chamber dimensions. The roof was changed from wedge-shaped into elliptical.

The aim was to obtain stress extremes in the underground excavation of chambers and when approaching workings of the opencast. If a chamber is filled dry with the size of pieces up to 200 mm, it has an about 25 m high (30% of the chamber height) cavity after filling shrinkage, the prospect of a dry filling was estimated when combining underground and open methods of deposit mining.

The results of simulation are as follows.

Projection on a vertical plane:

1. For a lateral distance factor λ =0.5, the maximum stresses in lock zones of roof and chambers walls are: 75.6x7.5 = 567 kgs/cm^2. At the top of the roof arch the stress is 75.6x2.0 = 151.2 kgs/cm^2. In the inter-chamber pillar the maximum compressive stresses decrease to 75.6x6.5 = 491.4 kgs/cm^2.

2. For a lateral distance factor λ =1.0: the maximum stresses in lock zones of roof and chambers walls are: 75.6x6.5=491.4 kgs/cm^2; being the same at the top of the roof arch; in the inter-chamber pillar the maximum stresses decrease to 75.6x5.5=415.8 kgs/cm^2.

3. For a lateral distance factor $\lambda = 1.5$: the maximum stresses in lock zones of roof and chambers walls are 75.6x6.5 = 491 kgs/cm^2, at the top of the roof arch compressive stresses increase up to 75.6x 8.5 = 642.6 kgs/cm^2.

Vertical section of the chamber:

The zones of maximum stresses are roof arch and bottom trench.

For coefficient of lateral extension - 0.5; 1.0; 1.5 the stress is 415.98; 1020.6; 1400 kgs / cm^2, respectively.

With regard to simulation results it was concluded that configuration and orientation of cavities affect the stability of combined excavations in the zone. A high concentration of stresses is revealed in the zone of the bottom trench entry provided that it has a wedge shape.

The spherical arch provides approximately twofold reduction in stress. The 30 m wide inter-chamber pillar provides stability, when decreasing width of the inter-chamber pillar to 20 m its stability drops. The maximum concentration of stresses is registered in the zone of the trench entry.

The problem of inter-chamber pillar mining optimization was solved with the help of modeling. Modeling conditions:

- lateral extension is 0.5H; 1.0H; 1.5H; 2.0H;
- inclination angle of the main force vector to the vertical axis $\alpha = 0^0$; 30^0; 40^0; 50^0 respectively for each lateral extension;
- Variants without filling both chambers and with filling of one chamber. The filling module is E = 0.1 MPa, that of the field is E = 1.4 MPa, 14 times higher than filling module.

Stress in any point of the model:

$$\sigma_m = \sigma^{1,0} n^0 \quad (6)$$

where $\sigma^{1,0}$=0.1 kgs/cm² for one stripe; n^0 - stripe number in the model point.

Natural stress:

$$\sigma_H = \gamma H \frac{\sigma_m}{\Sigma}, \quad (7)$$

where $\gamma = 3.0$ t/m³ – density of ore and rocks; H = 30 kgs/cm² -depth of occurrence from the day surface.

Analyzing the simulation results it can be concluded (Fig. 3):

- chamber roofs and bottoms are most stressed;
- arch-formed roof reduces the stress by 25%;
- filling of one chamber reduces the stress in the roof top by 2 times;
- in the inter-chamber pillar increased stresses appear.

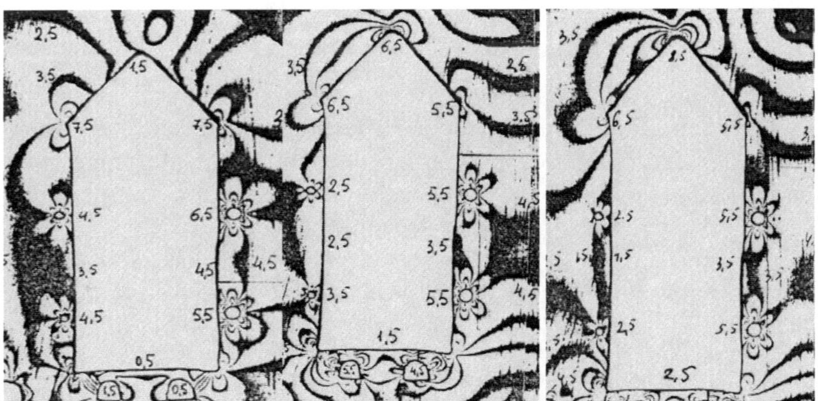

Fig. 3. Stresses in the elements of the open chamber vs. horizontal stress: 1-0.5; 2- 1.0; 3-1.5.

4. Results and discussion

Open-underground mining of deposits is difficult to control in terms of geo-mechanical field stability. Dry cavity filling has no effect on the situation [7-8].

Under equal conditions the reaction of field to technological impact depends on magnitude and sign of stresses in the field. Field displacement is a consequence of internal deformation and destruction of elementary mineral particles, complicated by medium anisotropy [9-11].

The internal stress field of structural separate parts of the deposit depends on the external stress field:

$$\sigma_{b.} = [\ \sigma_{MS}\, f\, \Sigma(\ 1...n\ )\ T]\ k_{gn} \tag{8}$$

where σ_b is the stress field of rocks; σ_{ms} – massif stress field; T – tensors of elastic inhomogeneity moduli: random, constant, etc.; k_{gn} – coefficient of geological disturbance.

In volumetric compression conditions in massifs, tectonic structures close, strength and modulus of deformation increase, as a result, the volumetric stress state transformations appear; stretching and shear on the cavity edge to compression at the depth of the mass. In the short-term destruction process in the out-of-the-way stage rheological processes take place in the out-of the limit area.

Depending on rocks deformation of in the massif, there are several destruction areas simultaneously [12-13]. Near the rupture there is a zone of broken rocks with a minimal supporting capacity. It is followed by weakened rocks of transcendent deformation, the strength increases as far from the rupture. Finally, there is a zone of untouched rocks in the stage of pre-limit deformation. Maximal preservation of the massif is possible via minimizing the first two zones and increasing pre-limit deformation, where strength is determined in terms of time:

$$\sigma ci\ (t) = kt \times [\sigma cl + (\sigma ci - \sigma cl\ )]\ e - at \tag{9}$$

where $\sigma_{ci.}$ – limit of instantaneous strength at one-axial compression, MPa; σ_{cl} – limit of the long-termed strength at one-axial compression, MPa; K_t – tectonic faulting coefficient; a – approximation parameter; t - time.

Condition of massif stabilty:

$$\Sigma\ (\ \sigma_3 + K\ \sigma_2\ \sigma_1\ ) \leq \Sigma\ \sigma_o\ K \leq \Sigma\ \sigma_н\ K \tag{10}$$

$$\Sigma\ \sigma_н\ \ K = f\ (\ \sigma_r\ ,\ h_c,\ P_{min}\ ,\ P_{max}\ ) \tag{11}$$

$$\Sigma\ \sigma_o\ K\ K_{pн} = f\ (\ \sigma_н\ ,\ H\ ,\ B\ ) \tag{12}$$

where $\sigma_{3,2,1}$ – main stresses in the massif; $\sigma_н$, σ_o – stresses in the zones of destructed and weakened rocks, respectively, MPa; h_c – height of the rupture affecting zone, m; P_{max} and P_{min} – maximal and minimal technological impact; σ_r – residual strength of rocks at one-axial compression, MPa; H and B – height and width of technologically affecting zone, m; K - factor of tectonic faulting impact; $K_{pн}$ – factor of stress loading out.

The control of massif state is in meeting the following conditions:

$$[\sigma_{HM}] < \sigma^R < [\sigma^R]_C \tag{13}$$

where $[\sigma^R]_C$ - critical stresses in the destructed rocks.

Zone height of critical stresses propagation:

$$H = f[\sigma]_C^{\ R} > h_T = f[\sigma]_T \tag{14}$$

where hT – zone height of technology-related impact, m;

$$h_T = \int_{min}^{max} fV \int_{min}^{max} fS \int_{min}^{max} fm \tag{15}$$

where H' – zone height of deformation to critical stresses, m; S_p - area of weakening, m²; α_H - dip angle, degree; m_n – normal thickness, m.

Geo-mechanical processes of rocks destruction and technological indexes are expressed in the condition [14-15]:

$$\sigma \cdot K_3 = \int_{l\min}^{l\max} fx\left(d_{x1}, d_{x2} ....d_{xn}\right) \rightarrow P, R = \int_{l\min}^{l\max} f \cdot x\left(d_{hZ} + dh_p\right) \quad (16)$$

where σ - stresses in the mining-affected zone, MPa; K_3 – coefficient of stress correction; l_{max}, l_{min} – outcropping spaces of rocks, m; $x_1...x_n$ – technological, physical and mechanical and other characteristics; P – ore losses, share units; R – ore contamination, share units; $h_з$ – height if the filled massif, m; h_P – height of mining effect, m.

The proposed model is special because of coefficient of stress correction K_3. In filling cavities with hardening mixtures the level of stresses is reduced due to characteristics of filling hardening material (Fig. 4).

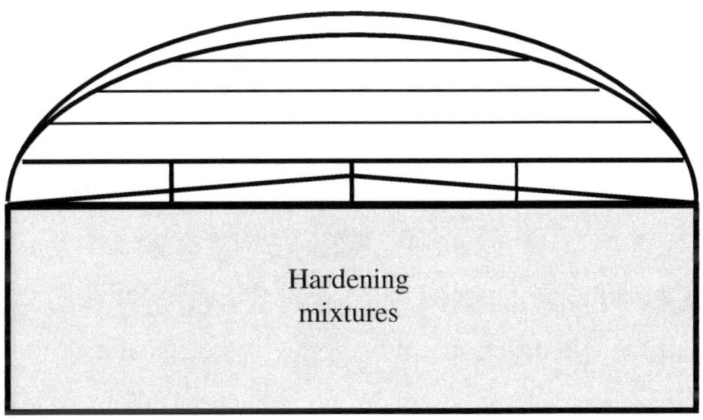

Fig. 4. The layout of roof support with hardening mixtures to support the ore-hosting layer

Work performance depends on rock mass vs. side support of the artificial massif walls. In conditions of volumetric compression, filling strength increases 1.5-3 times. The state of massifs, losses and contamination are determined by the volume of cavities, the volume of mined out ore and strength of rocks.

As for S.V. Vetrov massif stability is provided according to the condition:

$$L_P > 2\alpha = d_1\left(\frac{R_c}{KH\gamma} - 1\right), \quad (17)$$

where a - is a half-space of the blocked up rocks, m; d_1 - horizontal size of the structural block of rocks, m; R_c - rocks resistance of the structural block to compression, t/m²; γ - volume weight of rocks, kg/m³; H_γ- depth of the blocked rock arch shoe, m; K - safety factor.

Filling of cavities with hardening mixtures stabilizes the massif state; the effect is possible due to reducing losses and contamination. The possibility of stress control is provided by limiting deformations in filling cavities with hardening mixtures:

$$\sigma_1 \leq \sigma_2 \leq \sigma_3 = \sigma_H \cdot K_1 K_2 K_3 K_4 \tag{18}$$

where σ_1 – stresses in the zone of undamaged rocks, MPa; σ_2 - stresses in the clean-up operation zone, MPa; σ_3 - stresses in the filled massif, MPa; σ_H - rated resistance to compression of filling, MPa; K_1 - coefficient of filled massif inhomogeneity; K_2 - coefficient of increasing strength upon time; K_3 - coefficient of increasing filling strength in the massif; K_4 - of work conditions.

Stresses after filling the cavities:

$$\sigma_{..} = n_1\sigma_{n..3} + n_2\sigma_{c..3} + n_3\sigma_{m..3} + n_4\sigma_{H..3} + n_5\sigma^R = \sum_1^{\ell} n_\ell \sigma_m^y \tag{19}$$

where $\sigma_{n.3}, \sigma_{c.3}, \sigma_{m.3}, \sigma_{H..3}$ - support values of high-, medium- and low-strength hardening mixtures; ℓ - number of hardening elements; $n_1, \ldots, n_5$ - mass number of material in the amount of mixture; σ_m^y - strength of mixtures.

Model of controlling the massif state via stress uploading:

$$\sigma_1 \leq \sigma_2 \leq \sigma_3 = \sigma_H \cdot K_1 K_2 K_3 K_4 \tag{20}$$

$$PY_T = C_T - Z_T - Y_{T_{\alpha T}} - Y_{T_{RS}} - Y_{T_{RP}} \tag{21}$$

$$B_a = B_c E_B , \tag{22}$$

where σ_1 – stresses in the zone of undamaged rocks, MPa; σ_2 - stresses in the clean-up operation zone, MPa; σ_3 - stresses in the filled massif, MPa; σ_H - rated resistance to compression of filling, MPa; K_1 - coefficient of filled massif inhomogeneity; K_2 - coefficient of increasing strength upon time; K_3 - coefficient of increasing filling strength in the massif; K_4 - of work conditions; PY_T - profit from ore excavation and processing, ruble/t; C_T value of the mined out ore, ruble/t; Z_T full costs of ore mining and processing, ruble/t; $Y_{T_{RS}}$ – loss caused by 1t of contaminating mass along the block edge, ruble/t; $Y_{T_{RP}}$ – loss caused by processing of 1 t of contaminating mass inside the block, ruble/t.; B_a – number of alternative binding agents, weight units; B_c – number of standard binding agents, weight units; E_B – coefficient of binder equivalence.

The possibility of control stresses makes it possible to use low-strength compositions of hardening mixtures based on recycled ore dressing wastes after activation, e.g. mechanic and chemical technology for filling the mined out zones of the massif [16-17].

5. Conclusions

Performance indicators in underground mining of multi-structural fields can be improved via using possibilities of ore-hosting fields to develop supporting constructions, which keep the balance of the system "surface – massifs – environment".

The relation of natural to anthropogenic factors can make more precise the standards of developed, prepared and ready to excavation ore reserves relying on computational methods. Economic efficiency can be improved via better control of mining operations.

Development of a database containing information on the deposit and optimization of performance indexes with the help of geo-mechanical recourses of ore-hosting fields provide a balance of geo-mechanical system with low economic expenses and minimal losses and contamination of ores at the stage of mining processing. It is possible to minimize critical stresses and corresponding deformations due to dividing the ore field into sectors safe from the standpoint of geo-mechanics, and using less cost-demanding ways of keeping rock massif stable, as well as to specify norms of mined out, prepared and ready for excavation ores.

References

[1] Golik V.I. Prirodoohrannye tehnologii razrabotki rudnyh mestorozhdenij (in Russian) *Environmentally friendly technologies of ore fields development* // Moscow, INFRA – M, 2014. Series. Vysshee obrazovanie: Bakalavriat, 192 p.

[2] Liang Z. Z., Xing H., Wang S. Y. et al. A three-dimensional numerical investigation of the fracture of rock specimens containing a pre-existing surface flaw // Computers and Geotechnics. 2012. Vol. 45. P. 19–33.

[3] Vetrov S.V. Dopustimye razmery obnazhenij gornyh porod pri podzemnoj razrabotke rud (in Russian) *Permitted rock outcrop in underground ore mining* D. M., Nauka, 1975.270 p.

[4] Golik V.I. Special'nye sposoby razrabotki mestorozhdenij (in Russian) *Special ways of field development* // Moscow, 2014. Series. Vysshee obrazovanie: Bakalavriat, 132 p.

[5] Ismailov T.T., Jezdekov M.V. Koncepcija kombinirovanija geotehnologij (in Russian) *Concept of combined technologies*. Mining informational and analytical bulletin. M. 2006. No 8. p.23.

[6] Golik V.I., Poluhin O.N., Petin A.N., Komashhenko V.I. Jekologicheskie problemy razrabotki rudnyh mestorozhdenij kma (in Russian) *Ecological problems of KMA ore-hosting fields development* // Mining journal. 2013. № 4. C. 91-94.

[7] Komashhenko V.I., Vasil'ev P.V., Maslennikov S.A. Tehnologijam podzemnoj razrabotki mestorozhdenij KMA nadezhnuju syr'evuju osnovu (in Russian) *Reliable resource base to the technologies of KMA fields underground mining*. Izvestija Tulskogo gosudarstvennogo universiteta. Sciences of Earth. 2016. No 2. pp. 101-114.

[8] Kotenko E.A, Golik V.I., Hadonov Z.M. Upravlenie tehnologicheskimi kompleksami pri razrabotke rudnyh mestorozhdenij (in Russian). *Control of technological facilities in developing ore-hosting fields*. Vladikavkaz: Terek, 2000. – 290 p.

[9] Zoback M. L., Zoback M. D., Adams J. Global patterns of tectonic stress nature // Nature. 1989. Vol. 341. No. 6240. P. 291–298.

[10] Ljashenko V.I. Prirodoohrannye tehnologii osvoenija slozhnostrukturnyh mestorozhdenij poleznyh iskopaemyh (in Russian) *Nature-saving technologies in development of mineral deposits with complex structure* // Mine Surveying Bulletin. – 2015. – No 1. – pp. 10–15.

[11] Pagiev K.H., Golik V.I., Gabaraev O.Z. Naukoemkie tehnologii dobychi i pererabotki rud (in Russian) *Science-based technologies in ore mining and processing* // North-Caucasus Mining and Metallurgical Institute (State Technological University). Vladikavkaz, 1998, 571 p.

[12] Vorob'ev A.E., Razorenov Ju.I., Ignatov V.N., Dzhimieva R.B. Innovacionnye geotehnologii razrabotki mestorozhdenij gorjuchego slanca i vysokovjazkoj nefti (in Russian) *Innovative geo-technologies of oil shale and highly viscous oil fields developmen*t //Study guide for master students majoring in mining-geology and oil-exploration / Novocherkask, 2008, 214 p.

[13] Shestakov V.A., Razorenov Ju.I., Gabaraev O.Z. Upravlenie kachestvom produkcii na gornyh predprijatijah (in Russian) *Control of product quality at mining enterprises* // Recommended by Academic Council in Mining Education of the Ministry of Education, the Russian Federation as a teaching aid for University students, majoring in specialty 090200 / Novocherkask, 2001, 262 p.

[14] Stradanchenko S.G., Shinkar' D.I., Maslennikov S.A. Sostojanie i perspektivy razvitija kreplenija vertikal'nyh stvolov v slozhnyh gorno-geologicheskih uslovijah (in Russian) *State of the art and prospects of developing vertical columns fastening in difficult mining and geological conditions* // Mining informational and analytical bulletin. – No 2. – M. 2013.– pp. 26-34.

[15] Ilimbetov A.F., Ryl'nikova M.V., Radchenko D.N., Milkin D.A. Novye reshenija problemy kompleksnogo osvoenija rudnyh mestorozhdenij (in Russian) *New solutions to the problem of complex developing ore-hosting fields*. Vestnik of Nosov Magnitogorsk State Technical University 2006. No 4. pp. 8-13.

[16] Kidybinski A. The role of geo-mechanical modelling in solving problems of safety and effectiveness of mining production // Archives of Mining Sciences. 2010. Vol. 55. No. 2. P. 263–278.

[17] Sun Chao, Bo Jing-shan, Liu Hong-shuai. Study on Influencing Factors of Ground Settlement over Mined–Out Area // Journal of Jilin University. – Earth Science Edition. – 2009. –№3. – p. 498–502.

IOP Conf. Series: Materials Science and Engineering **221** (2017) 012014 doi:10.1088/1757-899X/221/1/012014

The Structure of Reclaiming Warehouse of Minerals at Open-Cut Mines with the Use Combined Transport

D A Ikonnikov[a], S V Kovshov [b]

Mining Department, Saint Petersburg Mining University, 21 line, Vasilevsky Island, 199106, Saint Petersburg, Russian Federation

E-mail: [a]ArtSciencesD@gmail.com, [b]KovshovSV@mail.ru

Abstract.
In the article performed an analysis of ore reclaiming and overloading point characteristics at modern opencast mines. Ore reclaiming represents the most effective way of stability support of power-intensive and expensive technological dressing process, and, consequently, of maintenance of the optimal production and set-up parameters of extraction and quality of finished product. The paper proposed the construction of the warehouse describing the technology of its creation. Equipment used for the warehouse described in detail. All stages of development and operation was shown. Advantages and disadvantages of using mechanical shovel excavator and hydraulic excavator "backdigger" as a reloading and reclaiming equipment was compared. Ore reclaiming and overloading point construction at cyclical and continuous method of mining using a hydraulic excavator "backdigger" was proposed.

1. Introduction.

The effectiveness of the solution of the production quality and growth of technical and economic indicators of ore-dressing enterprises depends on achievement of chemical and ingoing material composition stability - ores. Technological indicators of ore-dressing increase if ore arrives homogeneous in composition to a dressing factory during the long period. It allows to apply and accurately maintain the optimal production cycle chosen for it. Besides, it leads to increase in productivity of dressing sections, concentrate yield and extraction of the useful component into the concentrate, to reduce a losses in reject, to reduce an amount of reagent, to reduce the burden on the environment [8], to improve a concentrate quality, etc.

Increasing with the depth of opencast mining is a preservation problem of high mining equipment productivity arises [2, 10]. For this propose combined transport systems can be used at opencasts. Which can be used with a view to operate the equipment at peak performance and minimal operating costs [1,3,4,5,13]. The connecting link of cyclical and continuous parts of the flowsheet process is the intermediates ore overloading point. It also performs the functions of ore overloading point and buffer capacity. Using new technology - mobile crushing plants and high-angle conveyor complex is expanding capability of cyclically-stream transport [9].

All advantages of using fullest combined transport scheme can be achieve by correct substantiation of its application areas and providing the necessary consistency level of product.

Improving the efficiency of traffic flows delivery mineral subject to the necessary organization averaged-dock warehouse is an important task in these conditions. The paper proposed the construction of the warehouse [7] describing the technology of its creation. Ore reclaiming represents the most effective way of stability support of power-intensive and expensive technological dressing process, and, consequently, of maintenance of the optimal production and set-up parameters of extraction and quality of finished product [1,2,11,12].

2. Technology of transportation.

Schematic diagram of a reclaiming for the cyclic and continuous technology of transportation at opencast shown in figure. 1.

The most distribution on Russians opencasts has backhoes reloading points with application a mechanical shovel excavator. This may be due by the following reasons: the possibility of receiving and shipping large volumes of rock mass, independence of an equipment, simplicity of the device, high speed of construction, the ability to sort and ore reclaiming. However, there are certain disadvantages excavator overloads that reduce the effectiveness use combined transport system and limiting depth of its introduction into the opencast.

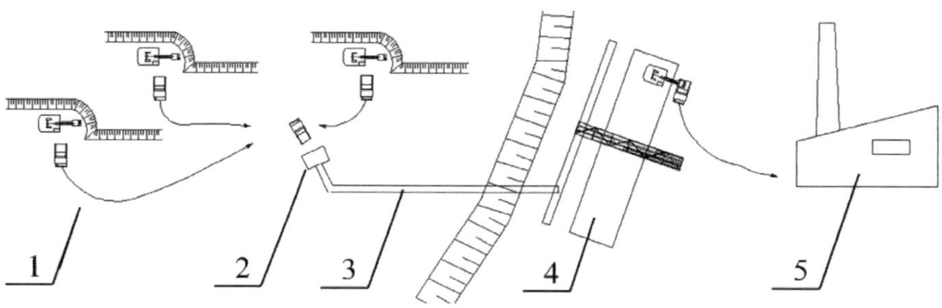

Figure 1. Schematic diagram of a reclaiming for the cyclic and continuous technology of transportation at opencast, where is: 1 – opencast mine; 2 - crushing plant to prepare for transportation by conveyor. 3 - conveyor system (high-angle conveyor system); 4 - ore reclaiming and overloading point (warehouse); 5 - factory.

An analysis of the scientific literature shows, that despite the widespread hydraulic excavator "backdigger" [14,15] in mining enterprises, the possibility of their use as a warehouse equipment been investigated bad. The hydraulic backhoes are universal. It is suitable for developing rocks, both above and below its level installation. The most efficient scheme with the lower digging and unloading of rock in means of transport standing below installation level of excavator. Savings are achieved by reducing angle of excavator rotation and lift height of excavator bucket.

General view of the warehouse with a location of roads [7] (as the third part of the transport system) is shown in figure. 2.

An ore is not located in the piles; the ore is located in a trench below installation level of excavator in this warehouse design. Preparation works at construction the warehouse consists in formation of a receiving trench. The receiving trench parameters are determined by the operating parameters handling excavator required warehousing capacity and free space available. Parameters of over the cross section (width and height) are fixed and determined by parameters of loading equipment. Length of the trench (L) is determined from the necessary warehouse storage capacity. From the conveyor (object 6 in figure 2) ore stream reloaded on mobile conveyor (object 7 in figure 2) [6] which is delivering the ore to transport bridge with flow divider. Mobile conveyor (object 7 in figure 2) completes platform (object 8 in figure 2) with the flow divider (flat-bottomed vibrating feeder). The division performs traffic separation by longitudinal trajectory of free falling, for the necessary number of flows (in this case it is three). Increasing the number of rolling away cones of rock leads to a decrease of the length rolling cone prism each of these separately. This in turn has decreases a segregation of rock and increases indicators of ore averaging. The platform (object 8 in figure 2) is also provides support for the pillar of the transport bridge (object 9 in figure 2). The second pillar of the transporter bridge also is mobile and has a drive for a move together with mobile

IOP Conf. Series: Materials Science and Engineering **221** (2017) 012014 doi:10.1088/1757-899X/221/1/012014

conveyor along the warehouse by a railroad tracks. This construction moves simultaneously with the delivery of ore from the opencast mine. Thus, the ore is stowed by layers over the warehouse's length. It looks like a layer cake. After filling warehouse area #1 (object 11 in figure 2) the whole structure (the mobile conveyor and the transporter bridge) is moved along the rails to the site #2 (object 12 in figure 2). The operation of warehouse is repeated on the warehouse site #2. Along with filling of warehouse area, shipment of the rock mass is processed.

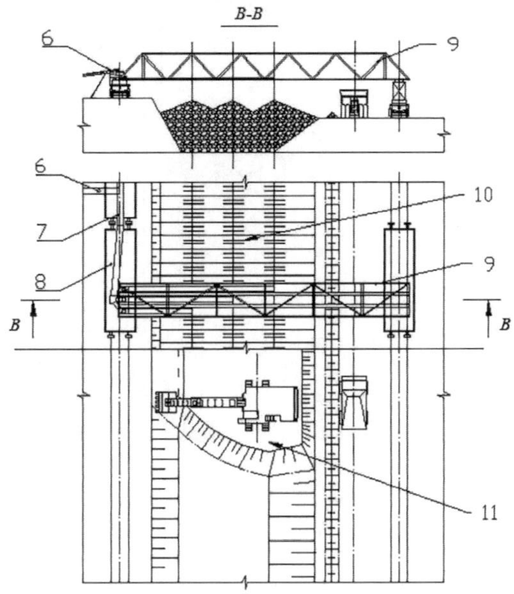

Figure 2. General view of the warehouse with a location of roads.
First of all, ore is loaded in a number faces of opencast in motor transport which deliver ore to ore preparation point (crushing plant (2)) for preparation for transportation by conveyor (figure 1). Further, the ore is transported by conveyor system (3) to the warehouse (4).

The generated surface of area #1 is leveling by bulldozer in order to create site for hydraulic excavator "backdigger". The process of ore averaging is carried out using a hydraulic excavator by cutting perpendicular to the layers of the "layer cake" warehouse. The averaged of rock mass is loaded by excavator in road transport (figure 3 a rail transport ore figure 4 a road transport).

IOP Conf. Series: Materials Science and Engineering **221** (2017) 012014 doi:10.1088/1757-899X/221/1/012014

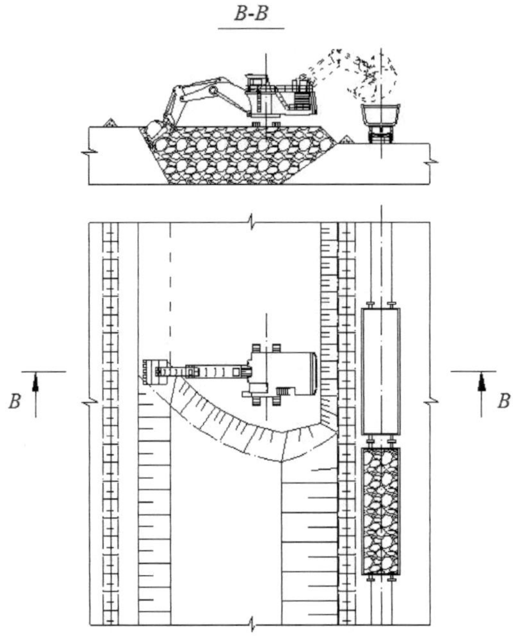

Figure 3. Loading operations in a rail transport.

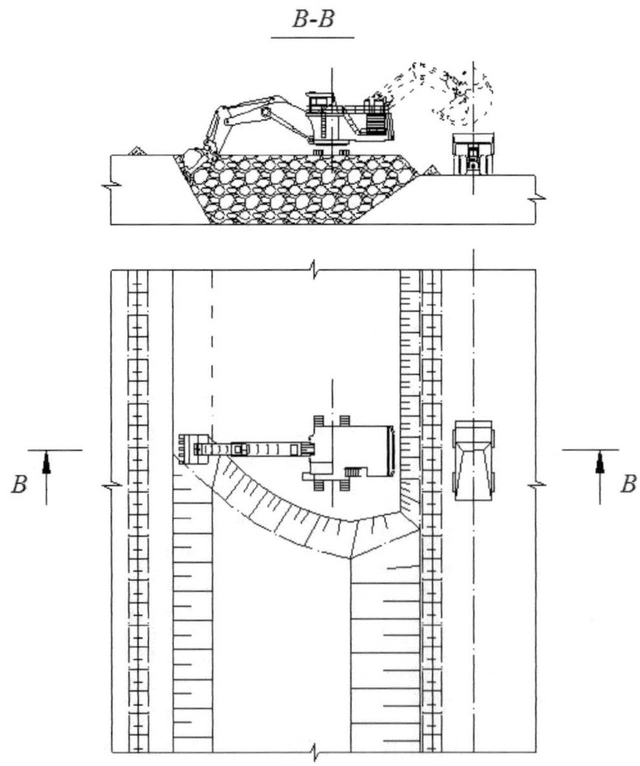

Figure 4. Loading operations in a road transport.

The parameters of warehouse's trench are shown in figure 5.

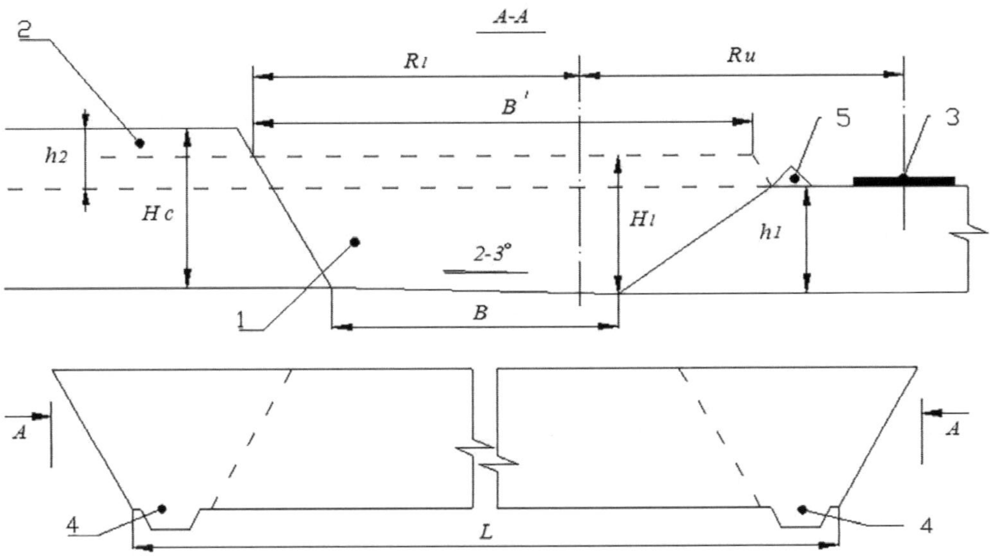

Figure 5. The parameters of warehouse's trench.

Initially, preparatory work should be carried out, which is consists in formation warehouse's trench (object 1 in figure 5) with a length equal to (L). The value (B) is the width of the trench at the bottom. The value (B') is the width of the trench at the surface. (B) and (B') is determined by the performance characteristics of the loading equipment. This is operating parameters of the excavator: digging reach at floor level (R_l), unloading radius (Ru), maximum digging depth (H_l).

Recoverable rocks that were extracted in the process of digging warehouse's trench is stowed along the top part one of the sides of the trench. Platform for mobile conveyor structure is formed of them (object 2 in figure 5). The total depth of the warehouse (Hc) is composed of platform height (h2) and depth of a trench (h1). Accordingly, solution to water drainage make bias towards the loading site at formation of the warehouse's trench. It support constant through excavating. At each end of the trench is formed water intakes (object 4 in figure 5), the parameters of it depend on the value of water inflow. Along the loading site of dump trucks (object 3 in figure 5) previously is formed safety shaft (object 5 in figure 5) to prevent a collapse of the rock mass.

3. Discussion and conclusions

The article suggests the design of the ore reclaiming and overloading point (warehouse) using a hydraulic excavator "backdigger" with application my own design formation of a stack of warehouse.

- Application of hydraulic excavator "backdigger" is rather perspective at ore reclaiming and overloading point. The hydraulic backhoes are universal and have many benefits.
- The formation of the ore reclaiming and overloading point of three cones (instead of a single) decreases segregation of rock particles by size. Consequently, division of ore chemical composition is decrease.

References

[1] Bastan P.P., Azbel E.I., and Klyuchkin E.I. The theory and practice ore averaging, Russian Federation, Moscow, Nedra (1979) P 255.
[2] Byzov V.F. Quality control of products at opencast mines, Russian Federation, Moscow, Nedra (1991) pp 161-227.
[3] Boguslavsky E.I. Quality management of ore: Textbook, Saint-Petersburg Mining University, Russian Federation, Saint Petersburg (2008) P 86.

IOP Conf. Series: Materials Science and Engineering **221** (2017) 012014 doi:10.1088/1757-899X/221/1/012014

[4] Ikonnikov D.A. Ore reclaiming process flowsheets for the cyclical and continuous method of opencast mining, Scientific Reports on Resource Issues 2011, Germany, Freiberg: TU Bergakademie, issue 1 (2011) pp. 79-82.
[5] Ikonnikov D.A. Formation technology of ore overloading and reclaiming on warehouse at application the cyclical and continuous method of transportation at open-cast mines, PhD thesis, National Mineral Resources University (University of Mines), Russian Federation, Saint-Petersburg. (2012) P 172.
[6] Kholodnyakov H.A., Ikonnikov D.A. and Argimbaev K.R. Mobile conveyor, Patent for a utility model RU #110724 U1. 2011.
[7] Kholodnyakov H.A. Ikonnikov D.A. Yakubovskiy M.M. et al. Method of ore transfer and reclaiming at stores in cyclic progressive opencut processes, Patent for an invention RU #2489339 C2. 2013.
[8] Koshkina V.S. Serova A.A. and Timofeev V.Y. Environmental-Toxicological Characteristics of Waters and Their Sources at Magnitogorsk With the Its Iron and Steel Industry. In IOP Conference Series: Materials Science and Engineering. IOP Publishing. Vol. 142 No. 1 (2016) p 012117.
[9] Loginov I.G., Slepjan V.I. and Malgin O.N. Design features of high-angle conveyor hoisting of hard rock in Muruntau opencasts mine, Russian Federation, Moscow, Mining magazine, issue 11 (2005) pp 63-65.
[10] Riko V.T., Chikota V.V., Kazachkov S.V., and Lomakin M.V. Reclaiming of multicomponent ores at opencasts mine, Russian Federation, Moscow, Mining magazine, issue 9 (2007) pp 45-48.
[11] Ryan P.J. Method and apparatus for accumulating stockpiles of flowable solid material, Patent for an invention US #4744459 B65G. 1988.
[12] Vajsberg L.A., Kruppa P.I. Perspective schemes of an ore dressing at cyclical and continuous method use, Russian Federation, Moscow, Mining magazine, issue 9 (2003) pp 30-33.
[13] Yakovlev V.L. Perspective solutions in the field of cyclical-continuous technology of deep open-pits, Moscow, Mining magazine, issue 4-5 (2003) pp 51-56.
[14] Yakubovskiy M.M. Application of hydraulic backhoe on reloading warehouses, Challehges and Solutions in Mineral Industry. Frieberger Forschungsforum 60. Berg- und Hüttenmannischer Tag., Germany, Freiberg: TU Bergakademie (2009) pp 130-133.
[15] Yakubovskiy M.M. Calculation of ore warehouses parameters by working out of complicated structure mineral deposits, Scientific Reports on Resource Issues 2010, Germany, Freiberg: TU Bergakademie, issue 3 (2010) pp 181-184.

IOP Conf. Series: Materials Science and Engineering **221** (2017) 012015 doi:10.1088/1757-899X/221/1/012015

Design of Engineering Surfaces Using Quartic Parabolas

A A Chekalin [a], M K Reshetnikov, V V Shpilev, S V Borodulina

Yuri Gagarin State Technical University of Saratov,
77 Polytechnique st., 410054, Saratov, Russia

E-mail: [a] graphic@sstu.ru

Abstract. The article considers the wireframe-kinematic method of hydrodynamic simulation of surfaces presented by the example of marine craft's bottom structure. The motion trajectory for the variable shape generator is normally set by a single or a number of directing lines. The law of shape variations of the generator is represented by the graphical or analytical display, or by a set of directing lines. All of these make up the so-called determinant of the surface, which includes the geometry component, or a certain initial position of the generator, and the algorithm component, i.e. the algorithm of the wireframe sealing. The directing line may be simple, described by a single equation, or compsite piecewise-smooth. In this situation, they are commonly modelled using splines, which results in piecewise-smooth surfaces.

1. Introduction

In mechanical engineering the majority of curved surfaces are referred to kinematic surfaces. This is due to the fact that when designing this type of surfaces, it is relatively easy to specify and provide their basic characteristics, and methods applied for their mechanical treatment are simple. Such surfaces can be used to solve the design and technological problems, e.g. the problems related with local shape control, calculation of mechanical characteristics, determination of cross sections and equidistance determination, etc. In certain instances, the engineering surface is designed as the wireframe, which is mainly in such industries as aircraft, shipbuilding and motor-vehicle construction, light and footwear industries. The frame of a surface is presented by the designer or is the result of scanning of the test model.

As is well-known, kinematic surfaces are formed by moving the constant or variable shape lines in accordance with a certain law [1, 7, 8]. Engineering geometry provides designers with a wide range of kinematic surfaces: spiral, cyclic, ruled surfaces with one, two and three directing lines, surfaces moved by parallel shifts and sections dependent surfaces, etc. The latter are frequently used in designing aircrafts and marine vessels.

In this paper we propose one of the wireframe-kinematic methods for hydrodynamic simulation of surfaces.

2. Research

The motion trajectory of the generator with a variable shape is generally set by one or several directing lines. The law of the generator's shape variations may be presented graphically or analytically, as well as by the directing lines. All these factors constitute the surface determinant. The latter is related with the geometrical component, i.e. a certain initial position of the generator, and the algorithmic

parameter, i.e. the algorithm for the frame sealing. The generators may be simple, which are described by a single equation, or composite piecewise-smooth. In the given instance they are generally modelled using the splines, and the surface can be described as piecewise-smooth.
Let us take the quartic curve for the generator, which presents a segment for interpolating and integrodifferential quartic spline (ID spline) [2, 6, 10]

$$\overline{r}(x,u)=\overline{r}_1(x)\phi_1(u)+\overline{r}_2(x)\phi_2(u)+\left(\overline{r}_1'(x)\phi_3(u)+\overline{r}_2'(x)\phi_2(u)\right)h+\overline{I}(x)\phi_5(u), \quad (1)$$

where $h=\dfrac{\left|\overline{r}_1'+\overline{r}_2'\right|}{2}$; $\phi_1(u)=(1+5u)(1-3u)(1-u)^2, \phi_2(u)=(3u-2)(6-5u)u^2$;

$\phi_3(u)=u(2-5u)(1-u)^2/2, \phi_4(u)=(1-u)(3-5u)u^2/2, \phi_1(u)=30u^2(1-u)^2$;

$\overline{r}_1$ and $\overline{r}_2$ are the radius-vectors of the endpoints of the arc; $\varphi(u)$ are the basic functions, and the parameter $u\in[0.1]$.

The vector $\overline{I}_i$ characterizes the curve deflection (Fig. 1). It highlights the centre of the nonplanar cross-section mass bounded by an arc and a chord. The vectors $\overline{r}_1'$ and $\overline{r}_2'$ are tangential.

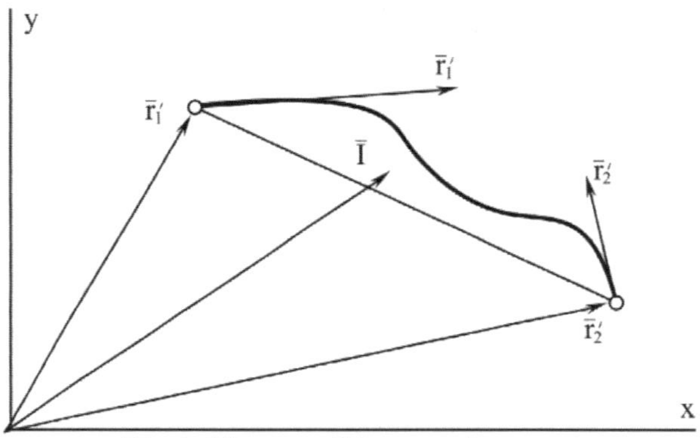

Fig.1. The arc of the quartic curve

We can see that the given curve has the parameters used to control the shape: two directing parameters at the endpoints of the arc and the vector-integral. In order to set the simple surface of dependent cross sections with a quartic curve used for the generator, it is necessary to set the directing lines for the endpoints of the arc, the generators for the tangential vectors, and the directing line for the vector-integral. As an example, let us consider the initial stage of hydrodynamic simulation of the surface of a small size hard-chine vessel. (Fig.2) The surface of the hull is designed as the kinematic surface of dependent plane sections.
Let the symmetry plane of a vessel coincide with the plane xOy. The axis of distances is parallel to the axis x, respectively. Then the dependent sections of the surface, or frames (which are also called lines), are parallel to the plane yOz. The surface of the hull (we consider only the left side) consists of two imperfectly joined parts: the side and the bottom. At the first stage, we set three directing factors: the side line, the bow and keel lines. The side line and the bow constitute the surface of the board side. When planning, the given surface is not washed by sea water and does not affect the quality characteristics of the vessel. Therefore, the surface is designed as the cylindroid, i.e. the surface with curved directing lines, the linear generator, and the parallelism plane yOz. The formula of the surface is given by $\overline{r}(x,v)=(\overline{r}_{sl}(x)-\overline{r}_{bw}(x))u$.
The bottom is constituted by the bow and keel. This part of the hull surface determines all the hydrodynamic parameters (Fig. 3).

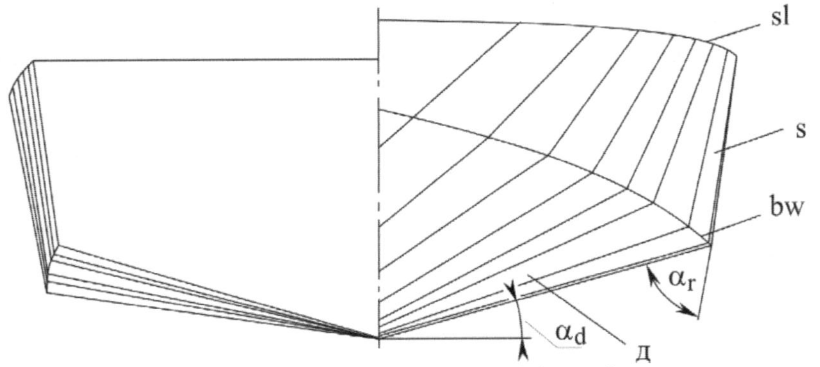

Fig.2. The bow-to-stern view of the surface
s – board side, b – bottom, sl - side line , bw – bow, αd - deadrise angle, αr - rebound angle

The directing lines are piecewise smooth. Each directing line is defined by twelve points. The point basis is interpolated by the spline. The cubic spline is commonly applied for the given purpose. As a result, we obtain two piecewise-smooth curves with the second order smoothness [2, 3]. At developing the surface, the generator (1) is shifted along these directing lines. To ensure the smoothness of the surface, it is necessary to provide the continuity and smoothness of the functions of cross-section parameters within the nodes of the directing lines.

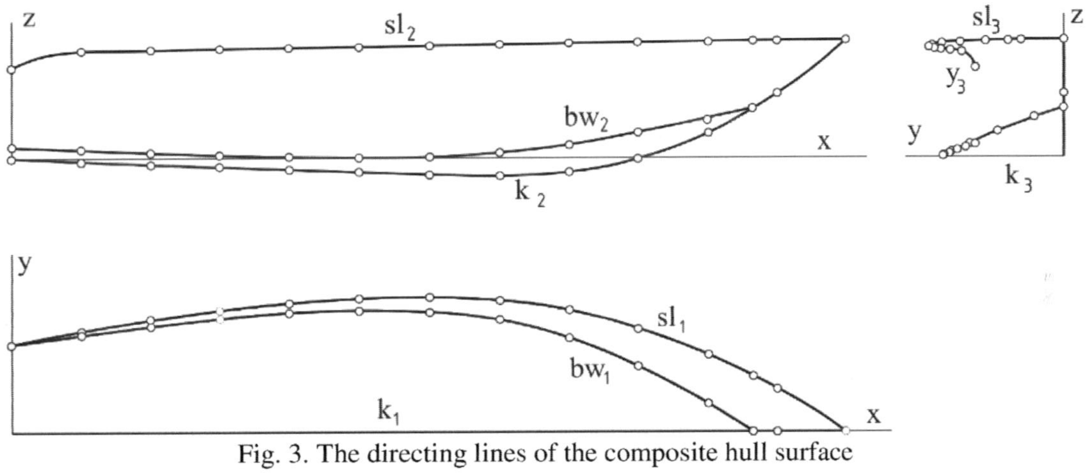

Fig. 3. The directing lines of the composite hull surface

For the cross-section surface we select the quartic parabola represented by the parametres (1)

$$\bar{r}(x,u)=\bar{r}^{\kappa}(x)\phi_1(u)+\bar{r}^{bw}(x)\phi_2(u)+\left(\bar{r}'^{\kappa}(x)\phi_3(u)+\bar{r}_4'^{bw}(x)\phi_2(u)\right)h+\bar{I}(x)\phi_5(u)\,, \qquad (2)$$

where $h=\dfrac{\left|\bar{r}'^{bw}+\bar{r}'^{\kappa}\right|}{2}$;

$\bar{r}_i^{\kappa}$ and $\bar{r}_i^{bw}$ and are the corresponding points on the keel and bow, having identical abscissas, i.e. they are on the single frame.

The indicated generator is characterized for five parameters: $\bar{r}^{\kappa}$ and $\bar{r}^{bw}$ are the endpoints of the arc found on the directing lines; $\bar{r}'^{\kappa}$ and $\bar{r}'^{bw}$ are the tangent vectors, which determine the deadrise angle α_d and the rebound angle α_r, respectively. The vector-integral $\bar{I}$ is the parameter, which ensures the control of the hull shape without altering the deadrise and the rebound. The given vector determines the area bound by the arc of the cross-section.

In order to develop the surface with the necessary shape, each frame must be presented as a straight line. To present the quartic curve in the form of the straight line, we need to set the parameters of the frame into initial positions along the entire length of the directing lines.
At the point $\overline{r}_i^{\kappa}$ we set the tangent vector $\overline{r}_i'^{\kappa}$ to the upcoming section, whereas at the point $\overline{r}_i^{bw}$ the tangent vector $\overline{r}_i'^{bw}$, i.e. the point of vector intersections, we shift to the middle of the chord $\overline{r}_i'^{\kappa} = \dfrac{\overline{r}_{bw} - \overline{r}_{\kappa}}{2}$, $\overline{r}_i'^{c} = \dfrac{\overline{r}_{\kappa} - \overline{r}_{bw}}{2}$.

Now we reduce the integral parameter to its original position. For this let us assign $\overline{I}_i = \dfrac{\overline{r}_i^{k} + \overline{r}_i^{bw}}{2}$, $i = 0,...13$.

We can now modify the wireframe surface, altering each frame separately and providing each $\overline{I}_i$ with an additional vector $\Delta\overline{I}_i$ and to the vectors $\overline{r}_i'^{\kappa}$ and $\overline{r}_i'^{c}$ we add the vectors $\Delta\overline{r}_i'^{\kappa}$ and $\Delta\overline{r}_i'^{c}$, respectively (Fig. 4).

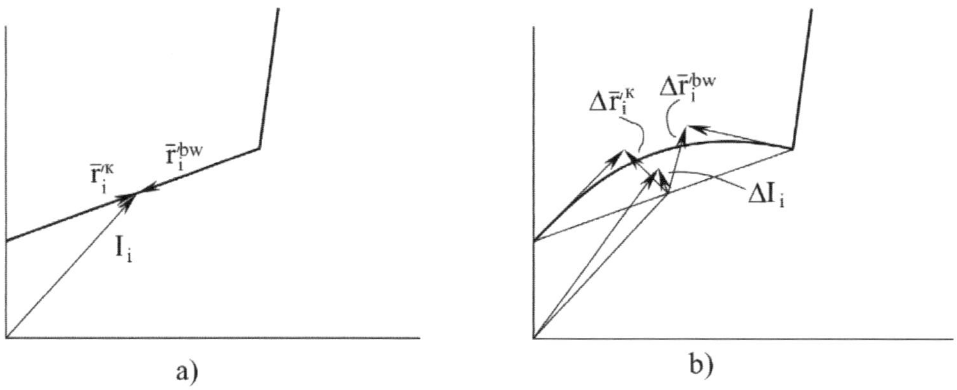

Fig. 4 Shape control of the i section
a) the original contour, b) the modified contour.

If we interpolate the values $\overline{I}_i$ by the parameter x, then we will not receive the ruled surface, since only the reference sections will have the straightforward form, and hence we need to interpolate the vector functions $\Delta\overline{I}_i$, $\Delta\overline{r}_i'^{\kappa}$ and $\Delta\overline{r}_i'^{bw}$. For this purpose, let us use the same polynomial as in the case of the longitudinal lines.
The results are presented graphically in the form of scalar functions, or the components of the vector function (Fig. 5).

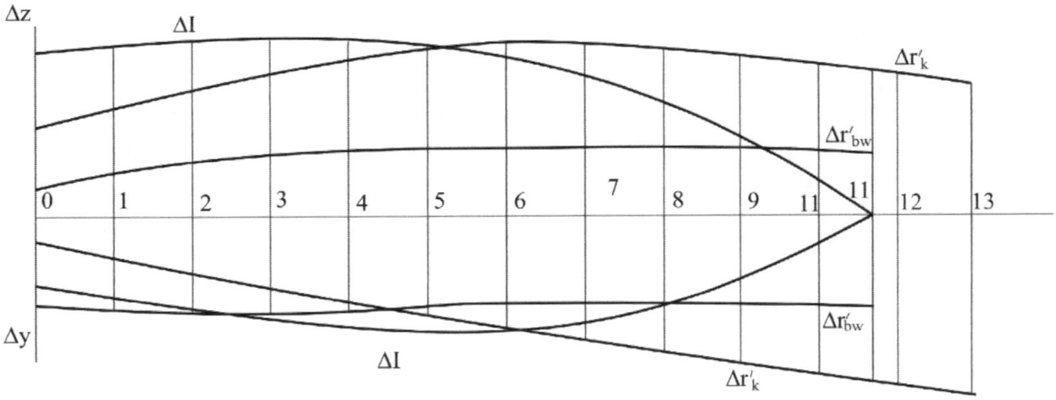

Fig. 5. The graphs for the quartic curve behaviour

Finally, the equation for the bottom surface takes the form

$$\bar{r}(x,u) = \bar{r}^{\kappa}(x)\phi_1(u) + \bar{r}^{bw}(x)\phi_2(u) + \left(\bar{r}'^{\kappa}(x) + \Delta\bar{r}'^{\kappa}(x)\phi_3(u) + \bar{r}_4'^{bw} + \Delta\bar{r}_4'^{bw}(x)\phi_4(u)\right)h + \bar{I}(x) + \Delta\bar{I}(x)\phi_5(u) \quad (3)$$

where $\bar{I}(x) = \dfrac{\bar{r}^{k}(x) + \bar{r}^{bw}(x)}{2}$, $\bar{r}_i'^{\kappa} = \dfrac{\bar{r}_{bw} - \bar{r}_{\kappa}}{2}$, $\bar{r}_i'^{bw} = \dfrac{\bar{r}_{k} - \bar{r}_{bw}}{2}$.

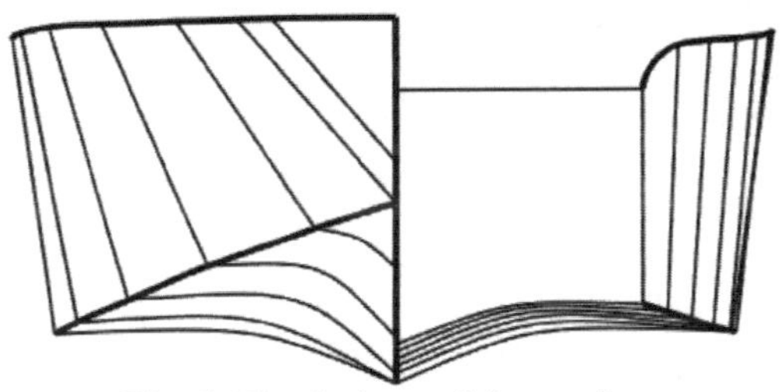

Fig.6. Final view of the surface

Figure 6 shows the final general view of the entire surface of the vessel.
It may be possible to smoothly join the side and the bottom, and thus make them into a single smooth surface.
Let us examine the surface of the hull of the round-bilge vessel (Fig.7).

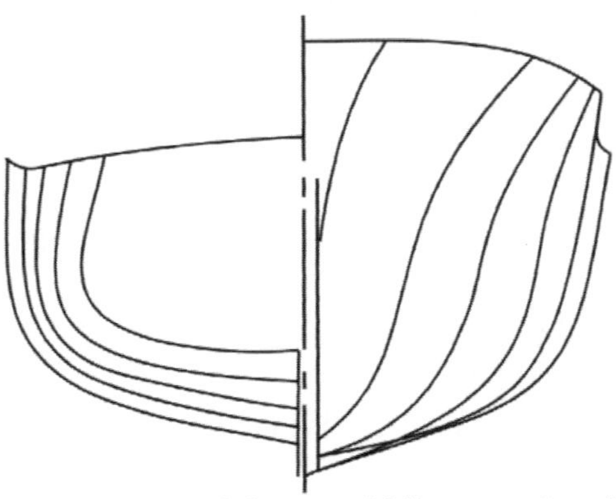

Fig. 7. The lines of the round-bilge vessel surface

IOP Conf. Series: Materials Science and Engineering **221** (2017) 012015 doi:10.1088/1757-899X/221/1/012015

As is seen from figure 7, the cross-section of the surface may have a double curvature and an inflection point. The given line may be presented as the cubic curve. However, let us make it composite for the accuracy and convenience purposes (Fig. 8).

If we have two component parts, then as in the previous case, we will have three directing lines. In this case, the inner directing line can be called considered as bow only as conditional assumption. As a component of the vessel design, it is missing.

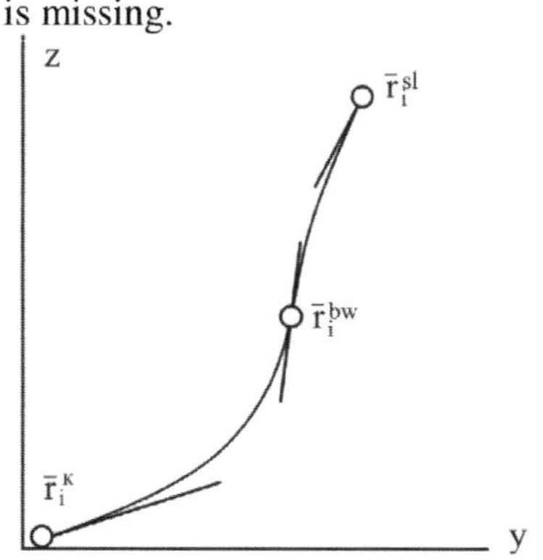

Fig. 8. The composite line of the i-th section.

It is recommended the points rc across all the sections be selected on the assumed points of the inflection, if there are any.

The line consists of two arcs of a quartic parabola (1). Let us write the equations to these arcs with respect to Fig.8:

$$\bar{r}_i^{b}(u)=\bar{r}_i^{\kappa}(x)\phi_1(u)+\bar{r}_i^{bw}(x)\phi_2(u)+\left(\bar{r}_i'^{\kappa}(x)\phi_3(u)+\bar{r}_i'^{bw}(x)\phi_4(u)\right)h_b+\bar{I}_i^{b}(x)\phi_5(u) \quad (4)$$

$$\bar{r}_i^{s}(u)=\bar{r}_i^{bw}(x)\phi_1(u)+\bar{r}_i^{sl}(x)\phi_2(u)+\left(\bar{r}_i'^{bw}(x)\phi_3(u)+\bar{r}_i'^{sl}(x)\phi_2(u)\right)h_s+\bar{I}_i^{s}(x)\phi_5(u) \quad (5)$$

where, $\bar{r}_i^{s}$ is the upper arc; $\bar{r}_i^{b}$ is the lower arc; $\bar{r}_i^{\kappa}$, $\bar{r}_i^{bw}$ and $\bar{r}_i^{sl}$ are the nodal points of the line; $\bar{r}_i'^{\kappa}$, $\bar{r}_i'^{bw}$ and $\bar{r}_i'^{sl}$ are the tangents to the line at the nodal points; $\phi_j(u)$, j=1…5 are the basis functions (1), $u\in[0.1]$ is the local parameter. The vectors $\bar{I}_i^{b}$ and $\bar{I}_i^{s}$ determine the deflection of the upper and lower arcs of the i-th section.

The design of the surfaces is to be conducted in line with the following rules:

The lines of the keel and the side are marked by the points. The points are interpolated by cubic splines [3, 4, 5, 9].

The cross-sections are pre-formed (Fig.7). A mid-point $\bar{r}_i^{bw}$ is set over each section.

The mid-points of sections are also interpolated by the cubic splines. A system of equations solved at constructing the spline, ensure smoothness for the curve being not lower than the second order [3]. However, we must be careful in order to make the bow line smooth without any oscillations or undesired points of inflection. Taking into account these requirements, the points shift their positions. It is also important that the first derived equations in the points $\bar{r}_i^{c}$ were equal left and right at each section.

The smooth vector functions $\bar{r}'^{\kappa}(x)$, $\bar{r}'^{bw}(x)$, $\bar{r}'^{sl}(x)$ $\bar{I}^{b}(x)$ and $\bar{I}^{s}(x)$ are composed and presented in the drawing by their scalar components (Fig.9).

To achieve the required characteristics for the engineering surface, its shape is changed by manipulating the functions $\bar{I}^{b}(x)$ and $\bar{I}^{s}(x)$.

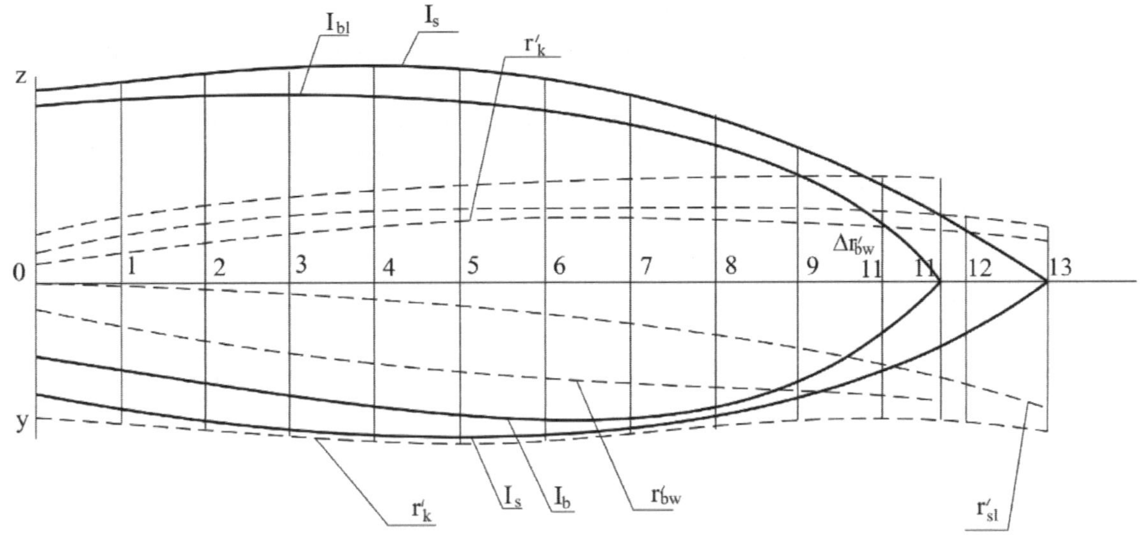

Fig.9. Functions determining the surface of a hull

The vectors $\overline{I}^b$ and $\overline{I}^s$ can be modified unrestricted, while maintaining the smoothness functions $\overline{I}^b(x)$ and $\overline{I}^s(x)$. Any intermediate cross-section will retain the first order smoothness which is sufficient for hydrodynamic properties of the surface.

Thus, we can design surfaces with cross sections composed of a larger number of arcs.

3. Conclusion

The above considered mathematical parameters allow for the design of complex engineering surfaces on the basis of dependent sections, and at the same time efficiently alter the shape of the surface locally. These alterations do not require any recalculation for the entire surface. The algorithms based on the given design method are simple and easy to program. Using this method, we can simulate the aero- and hydrodynamic properties of surfaces, complex surfaces of buildings and structures under the interactive mode, while adding appropriate tools to the existing computer-aided design systems.

References

[1] Ivanov G.S. Konstruirovanie tehnicheskih poverhnostej / G.S. Ivanov. M.: Mashinostroenie, 1982. 192 p.

[2] Chekalin A.A. Modelirovanie poverhnosti slozhnyh tehnicheskih poverhnostej na osnove integrodifferencial'nyh splajnov. Avtoreferat kandidatskoj dissertacii. 1998.

[3] Foks A., Pratt M. Vychislitel'naja geometrija. Primenenie v proektirovanii i na proizvodstve: Per. s angl. - M.: Mir, 1982. 304 p.,

[4] D. Rodzhers, Dzh. Adame. Matematicheskie osnovy mashinnoj grafiki. M., Mir, 2001, 605 p.

[5] Dergunov V.I., Zhilina N.D., Popov E.V. Osnovy komp'juternyh tehnologij v proektirovanii. – N.Novgorod: Izd-vo NNGASU, 2003. 157 p.

[6] Zamjatin, A. V. Algoritm approksimacii poverhnosti splajnami / A.V. Zamjatin, A.E. Kubarev, E.A. Zamjatina // «Naukovedenie», 2012, 3.

[7] Korotkij V.A. Komp'juternoe modelirovanie kinematicheskih poverhnostej / V.A. Korotkij, E.A. Usmanova, L.I. Hmarova // Geometrija i grafika. 2016. T. 3. . 4. P. 19-26.

[8] Golovanov N.N. Geometricheskoe modelirovanie: nauchnoe izdanie / N.N. Golovanov. M. Izd-vo fiziko-matematicheskoj literatury, 2012. 472 p.

IOP Conf. Series: Materials Science and Engineering **221** (2017) 012015 doi:10.1088/1757-899X/221/1/012015

[9] Zakharov O.V., Balaev A.F., Bochkarev A.P. Shaping of spherical surfaces on centerless superfinishing machines with longitudinal supply. Russian Engineering Research. 2015. 35. 4. P. 264-266.
[10] Barton M., Shi L., Kilian M., Wallner J., Pottmann H. Circular arc snakes and kinematic surface generation // Computer Graphics Forum "Eurographics". 2013. DOI:10.1111/cgf.12020

IOP Conf. Series: Materials Science and Engineering **221** (2017) 012016 doi:10.1088/1757-899X/221/1/012016

Machine-Tractor Aggregates Operation Assurance by Mobile Maintenance Teams

G V Redreev[1, a], O V Myalo[1, b], S P Prokopov[1, c], A P Solomkin[1, d], G A Okunev[2,e]

[1]Omsk State Agrarian University named after P.A. Stolypin, Russia, Omsk
644008, Omsk, Institutskaya Pl. st. 1, Tel. (8-3812) 650-172
[2]South Ural state Agrarian University, Russia, Chelyabinsk
454080, Chelyabinsk, Prospect Lenina, 75, Tel. (8 (351) 266-65-30

E-mail: [a]gv.redreev@omgau.org, [b]ov.myalo@omgau.org, [c]sp.prokopov@omgau.org, [d]ap.solomkin@omgau.org, [e] mail@csaa.ru

Abstract: operability of machine-tractor aggregates (MTA) is ensured by purposeful activity of maintenance and repair performers. MTAs operation assurance can provide achievement of absolutely different goals. For further development of technical service formation concept concretization in the part of determining locations of maintenance and repair performers and their area of expertise is suggested, as well as of arising peculiarities of equipment. The theoretical task is reduced to the type of tasks of distribution of recourses or transportation tasks. Mobile maintenance teams of regional agricultural equipment manufacturers' dealers have experience of technical service performance. The formed stream of requests from agricultural plants determines the direction of correcting of existing theoretical provisions, confirming necessity of further development of centralized method of technical service by efforts and means of manufacturers' dealers.

1. Introduction

Machine-tractor aggregates performance assurance can provide achievement of absolutely different goals determined by the type and composition of the aggregate, crop-growing production plan for the current year, maintenance and repair performer organizational relation, etc.

The most urgent operational goal is to ensure MTA reliability during performance of crop farming tasks. The strategic goal is to ensure MTA functioning during its standard service life.

Graphically the aims can be presented in the following way:

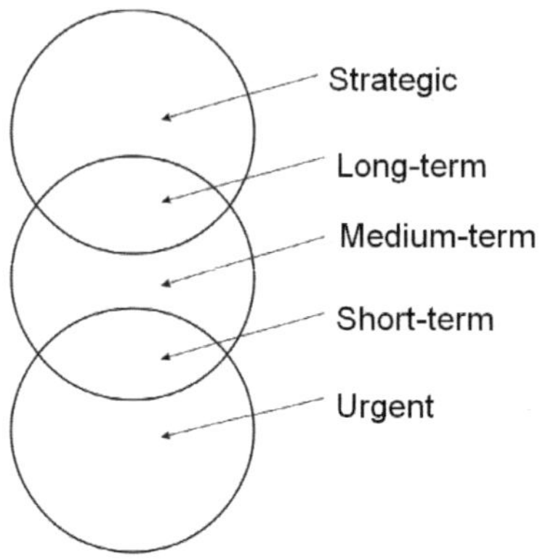

Picture 1. – Maintenance and repair aims

The goal choice can have significant influence on the applied maintenance and repair technologies and consequently on the quantitative and qualitative composition of maintenance and repair performer group.

We described the concept formation [1, 2] of MTA maintenance and repair process organization starting from the definition of basic notions, in the earlier published works [3]. The presented concept of the technical service suggests its further detailing in the part of location of the maintenance and repair performers and their field of expertise as well as arising peculiarities of equipment in disposal.

The technical service department organization structure represents separate maintenance and repair units of MTP in general and MTA in particular divided according to their number, location, equipment, and staffing level.

The regional-district division principle is based not only on the territorial adjacency but also on the similarity of agricultural production directions, adequate structure of MTP, similar staffing level.

Territorial independence is necessary for activity coordination of a designed body with local authorities represented by district administration for agriculture, State Technical Supervision Body inspections.

The structural organization can have several levels:

1– regional level
2 – district level
3 – inter-farm level
4 – plant level
with formed coordination and subordination connections.

The connections formation can take place as with creation of non-profit partnerships and associations, inter-farm specialized enterprises, as well as with formation of maintenance and repair of complex equipment affiliates by central organizations.

2. Research results

Let us consider the such basic notions as basics of developed concept, in philosophic aspect, as notions representing a model of thinking, "reflecting and fixating important traits of things and phenomena of objective reality". In the structure of the notion the dialectics of singularity, specific and universal finds reflection [4].

In accordance with triality of every notion it is possible to separate three levels of MTA operation assurance:

1-st level - monthly maintenance and maintenance-1 of MTA equipment and tractors.

2-nd level - complex maintenance of units, aggregates and MTA systems - motors, transmission, hydrosystems, electrical equipment, working parts.

3-rd level - periodic general control and instrumental evaluation of technical condition of MTA machines.

The first level can be performed by efforts of mobile maintenance teams of farms or by teams if inter-farm unions, i.e. on the inter-farm level and on the plant level.

The second level can be implemented by the efforts of mobile teams of special regional service and repair organizations, i.e. on inter-farm level.

The third level can be performed by efforts and means of dealers of agricultural equipment manufacturers i.e. on the regional level.

Thus, the task is to distribute the scope of MTA maintenance works between possible technical service providers of different levels.

To ensure MTA operation it is necessary to perform the volume of works Q_j of maintenance of different complexity x_{ij} (see the table) at known price for maintenance services of different level C_i,

Known:

- unit cost of crop loss a_{i1} at MTA downtime during performance of technical service of i level,
- index of mobile maintenance teams transportation costs b_{ij}.

Table 1. The distribution of the scope of MTA maintenance works between providers.

	Q_1	Q_2	Q_3
C_1	x_{11} b_{11} a_{11}	x_{12} b_{12} a_{12}	x_{13} b_{13} a_{13}
C_2	x_{21} b_{21} a_{21}	x_{22} b_{22} a_{22}	x_{23} b_{23} a_{23}
C_3	x_{31} b_{31} a_{31}	x_{32} b_{32} a_{32}	x_{33} b_{33} a_{33}

The task relates to the group of tasks on distribution of recourses or transportation tasks.

It is necessary to perform the scope of MTA maintenance works having distributed it across the levels of technical service in the following way:

$$Q_{ij} = \sum_{j=1}^{3} x_{ij} \,, \qquad (1)$$

With that the cost of expenses and losses must be minimal:

$$\sum_{i=1}^{3} \sum_{j=1}^{3} C_{ij} x_{ij} \, a_{ij} b_{ij} \rightarrow min \qquad (2)$$

Over the last years a new form of maintenance of agricultural plants became widespread - mobile maintenance teams that perform works on site. Mobile equipped maintenance teams perform works immediately at fields where tractors and self-propelled vehicles operate. This type of maintenance of MTA stock reveals and eliminates different failures as mobile team experts widely use means of technical diagnostics [5], which in its turn makes it possible to continue work in a short term.

As a rule mobile maintenance teams are formed at dealer centers which assume responsibility on organization and execution of warranty and post-warranty service in the scope and in volume specified by technical documentation [6].

Warranty service and dealer's maintenance [7] includes:

- providing and canceling equipment warranty registration;

IOP Conf. Series: Materials Science and Engineering **221** (2017) 012016 doi:10.1088/1757-899X/221/1/012016

- maintenance, replacement of defective details, equipment failure elimination on machines under warranty.
- instructing and training of farm machinery operators and agriculture experts
- control over following the operation rules by the owner, timely and full maintenance (monthly maintenance, maintenance-1, maintenance-2, conservation/starting after winter downtime);
- on-call consulting and on-site visit of technical center experts in order to eliminate technical problems
- supply control and equipment operation starting after winter downtime
- dealer's service in post-warranty period

Dealers perform a complex of necessary and accompanying measures on equipment maintenance under warranty. The scope and requirements to the service are defined in the "Technical specification and operation manual", Service book, and other instructions accompanying the equipment. [8].

Service center eliminates defects and recovers manufacturer's product during warranty period in the shortest possible time, but no longer than in three days starting from the date of receiving message about equipment failure. Warranty maintenance on manufacturer's recommendation is performed in the following periods [9]:

- if dismantling of main units is not required - during 24 hours;
- if dismantling of main units is required - 48 hours;
- dismantling is required together with replacement of basic details (frames, semi-frames, motor unit, drive axle, change gearbox) within reasonable time agreed with the contractor but no more than 72 hours from the moment of receiving from the manufacturer a message about equipment failure and equipment damage statement[10].

Manufacturers' dealers in the Omsk region are: supply base OAO "Semyrechenskaya", OOO "Omskdizel", OOO "Terra", OOO "AvtoSpetsMash", OOO "SeverTransAgro", supply base ZAO "Agromash", base AO "Agrokomplekt", base OOO "Sibirskaya", Federal Unitary Enterprise Omsk experimental plant, Private Entrepreneur Shumilov V.V., OOO "OmskAgroLizing", OOO "SibzavodAgro".

Let's consider the main trends of formation dealer's service in the Omsk region on the example of a dealer. Technical center of the supply base "Sibirskaya" performs not only warranty repair but also post-warranty period maintenance, as well as running repair of equipment purchased in other firms. The personnel accounts 14 people: chief engineers, mechanics, diagnostic personnel, setup technicians, fitters, fitters-electricians. Portative unit of maintenance is equipped with the following tools - diagnostic tools, set of wrenches, compression gauge, conditioning equipment charging tools [11]. Considering the distribution of maintenance requests among the districts of the Omsk region in 2016 (pic.2) one can note quite significant difference of requests frequency from zero to 40 requests for one season . When counting the number of requests to the service center, both types of requests were taken account - as for planned maintenance, and complaints about equipment failure among main customers of the supply base "Sibirskaya".

Data analyses allows for the conclusion that service is more required in districts located closest to Omsk as administrative center which is connected with lower transportation costs. Also maximum number of requests is registered in the districts where the largest agricultural enterprises with high turnover and revenue are located.

Agricultural enterprise where technical service and elimination of current defects (pic.3,4,5) were carried out are situated at different distances from the dealer center. In connection with this a decree was issued according to which tariffs for distance are as follows: up to 100 km in both directions costs 10 rub per 1 km, if the distance is more than 100 km, the tariff is 9 ruble per km.

Pic 2 shows that the farther an agricultural enterprise is from the dealer center the more expensive the warranty service is. In order to perfect the system of technical service it is needed to open additional warranty service points, it will cut losses on mobile maintenance teams servicing which in its turn also will make it possible to perform maintenance at shortest possible time.

IOP Conf. Series: Materials Science and Engineering **221** (2017) 012016 doi:10.1088/1757-899X/221/1/012016

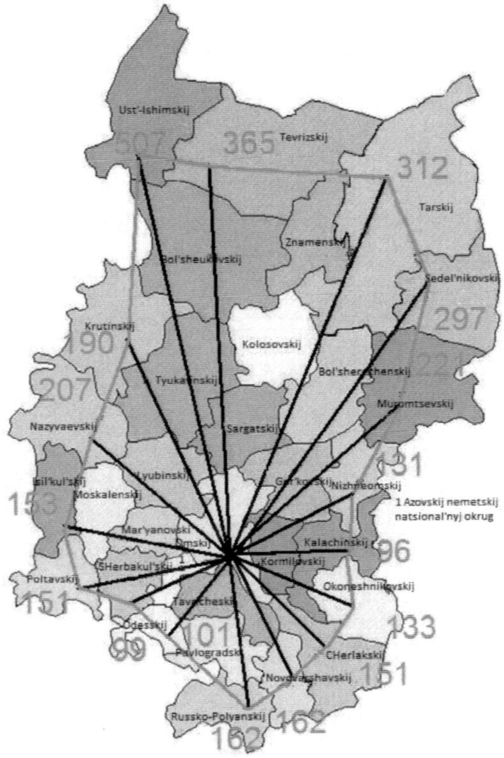

Picture 2. - Distribution of requests to the service center of OOO "Sibirskaya" supply base according to the districts of the Omsk region for the year 2016 [12]

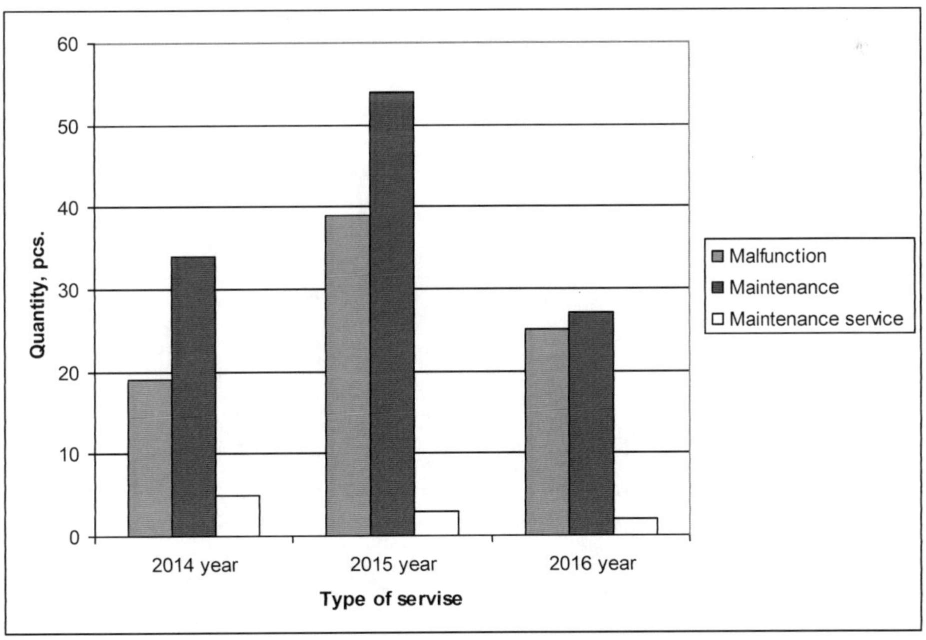

Picture 3. - Distribution of requests on Minsk Tractor Plants' tractors maintenance 82, 82.1, 82.2 by mobile teams of OOO "Sibirskaya" base (2014 – 2016) [12]

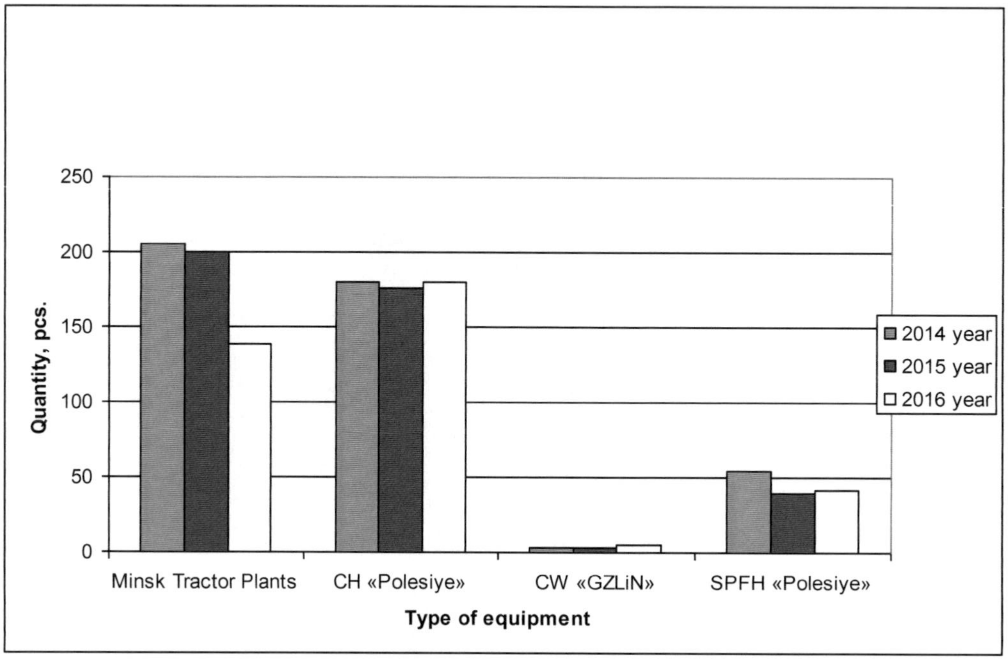

Picture 4. - Distribution of maintenance requests according to equipment type serviced by mobile maintenance teams of OOO "Sibirskaya" base (2014 – 2016) [12].

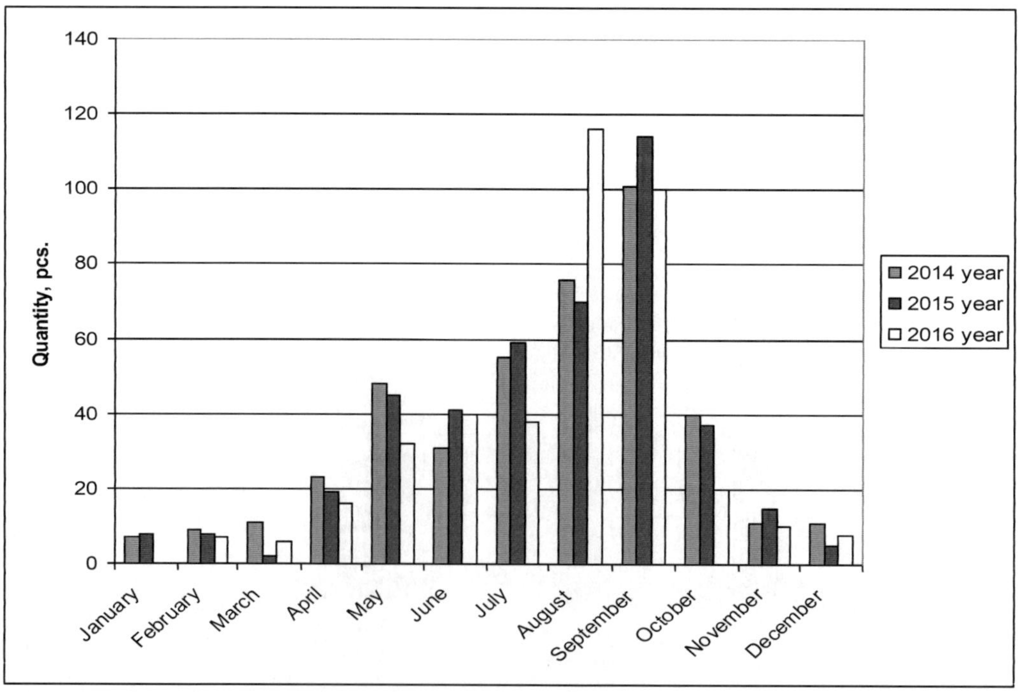

Picture 5. - Distribution of technical service requests for equipment maintenance performed by mobile teams of OOO "Sibirskaya baza" according to months (2014 – 2016) [12]

IOP Conf. Series: Materials Science and Engineering **221** (2017) 012016 doi:10.1088/1757-899X/221/1/012016

3. Conclusions

MTA operability assurance can help to achieve absolutely different aims, the most urgent is to ensure reliability of MTA during execution of current operations on crop growing. The choice of aim can have significant influence on applied maintenance and repair technologies and consequently on qualitative and quantitative composition of maintenance and repair performers. Organizational structure of service center represents regional independent maintenance and repair units of all MTP and MTA in particular.

Over the last years a new form of maintenance of agricultural plants became widespread - mobile maintenance teams that perform works on site of agricultural enterprises. As a rule mobile maintenance teams are formed in technical centers of dealers which undertake the responsibility on organization and performance of warranty period and post-warranty servicing of agricultural machinery in the scope and quality determined by technical documentation. Because of underdevelopment of engineering services in agricultural enterprises mobile maintenance teams also perform other works on servicing, maintenance and elimination of agricultural equipment failures.

Reference

[1] Conceptual thinking in finding solution of complex problems. / Teslinov A.G. -SPb.: Piter, 2009. 288 pages.

[2] Conceptual designing of complex solutions / Teslinov A.G. -SPb.: Piter, 2008. - 288 pages.

[3] Ensuring Machine and Tractor Aggregates Operability. / Redreev G.V. IOP Conference Series: Materials Science and Engineering, **142**, 1.

[4] Notion. / Voyshvillo E.K. - M., MSU, 1967. 288 pages.

[5] Complex system of maintenance and repair agricultural machines. (Part 1). GOSNITI, 1985. 144 pages

[6] Formation of technical service of agricultural equipment. / Babchenko L.A. Dissertation of Doctor of Sciences. Almaty, 2010. 567 pages.

[7] Basics of formation of systems of technical service in APK Sibir / A.E. Nemtsev, V.V. Korotkikh. Novosibirsk, 2009. 152 pages.

[8] Formation and assurance of tractors readiness / Solomkin A.P. Dissertation of Doctor of Science. Novosibirsk, 1989. 458 pages.

[9] Influence of wear factor on agricultural machines reliability / A.P. Solomkin, O.V. Myalo, S.P. Prokopov - Achievements of science and technology of agro-industrial sector. Moscow, 2015. **29**. 1. P. 61–63.

[10] Maintenance and repair of agricultural machines.: textbook / edit Chernoivanova V.I. - Moscow - Chelyabinsk: GOSNITI, CSAU. 2003. 992 pages.

[11] Organization of repair and maintenance service in agriculture sector: textbook / Yudin M. I., Stukopin N.I., Shirai O.G. Krasnodar: KSAU. 2002. 944 pages.

[12] Supply base "SIBIRSKAYA": Agriculture machinery in Omsk [electronic source, access mode: http://www.sibbaza.ru/]

IOP Conf. Series: Materials Science and Engineering **221** (2017) 012017 doi:10.1088/1757-899X/221/1/012017

Simulation of Roasting Metallurgical Concentrates in Fluidized Bed Using CFD-DEM

I I Beloglazov [a], Y V Kuskova

Saint-Petersburg Mining University, Mineral Processing Department
199106, St. Petersburg, Vasilevsky Island, 21 line 2, Russia

E-mail: [a] filosofem@mail.ru

Abstract. In this study, we utilized multiphase computational fluid dynamics (CFD), and discrete element method (DEM). Effect of the kinetic parameters of the roasting process in a fluidized bed was investigated. Our results indicate that it is possible to numerically integrate the coupled CFD-DEM system without significantly increasing computational overhead. It is also clear, however, that reactor operating conditions, reaction kinetics, and multiphase flow dynamics have major impacts on the roasting products exiting the reactor. We find that, with the same pre-exponential factors and mean activation energies, inclusion of distributed activation energies in the kinetics can shift the predicted average value of the exit gas-solid-phase and its statistical distribution, compared to single-valued activation-energy kinetics. These findings imply that accurate resolution of the reaction activation energy distributions will be important for optimizing roasting processes.

1. Introduction

Developing of coupling CFD-DEM models fluidized bed furnace requires careful consideration of the relevant physical processes: gas and particles dynamics of multiphase flow, the chemistry of the roasting process. The interaction between the phases must be properly taken into account for a realistic simulation. Laws that make up the mathematical formulation of each of these processes separately, are closely connected in a single system of equations[1].

Consideration of the various physical processes coupled together within one model, is quite a challenge for the simulation. On the other hand, this model will greatly improve the furnace roasting process by providing information on the desulfurization nickel concentrate, particle size distribution and compositions in the different zones of the furnace, in particular for the detection of high temperature zones, where there is excessive coarsening of the particles, resulting in the failure of the furnace.

Recently, computational models have provided considerable improvements due to high performance calculations using modern algorithms. A lot of CFD simulations using continuum models have been reported in literature [2,4,5]. However, these models show a serious over-prediction of the local heat transfer coefficient compared to experimental data and difficult formulation of constitutive equations (e.g. particle–particle drag) which may affect the simulation accuracy.

In this work, the multiphase phenomenon in liquid fluidized beds is studied by using the DEM-CFD method, which proved particularly promising [6,7] the fluid flow is obtained by solving the locally-averaged continuity and Navier-Stokes equations (CFD), while the motion of individual particles is studied by means of the Discrete Elements Method (DEM). This method presents the advantage of the capability to model liquids and particles properties and distribution [8].

There are mainly two typical multiphase CFD modeling techniques for studying fluidized bed processes, namely the Eulerian–Eulerian and Eulerian–Lagrangian approaches. In the Eulerian–Eulerian approach,

IOP Conf. Series: Materials Science and Engineering **221** (2017) 012017 doi:10.1088/1757-899X/221/1/012017

the two phases are treated mathematically as inter-penetrating continua and thus the approach is also called Two Fluid Model (TFM) [9]. Besides the fluid flow, an Eulerian framework is also applied to describe the motion of solid particles where the kinetic theory of granular flow (KTGF) is incorporated in most models to simulate particle collision for the closure of equation systems [9].

The advantage of the Eulerian–Eulerian approach is its superior computational convenience for multiphase flow. This gas–liquid–solid fluidized beds can be operated with different hydrodynamic regimes, which depend on the gas and liquid velocities, as well as the gas, liquid and solid properties. For proper reactor modelling, it is essential to know under which regime the reactor will be operating [10]. Two important hydrodynamic transitions within gas–liquid–solid fluidized beds are the minimum liquid fluidization velocity. The minimum liquid fluidization velocity is the superficial liquid velocity at which the bed becomes fluidized for a given superficial gas velocity; above the minimum liquid fluidization velocity, there is good contact between the gas, liquid and solid phases which is essential for heat and mass transfer processes [4,11,12].

2. Simulation models and methods

In the Eulerian–Eulerian approach, both gas and solid phases are treated mathematically as inter-penetrating continua. The basic idea is to use the governing equations correlating the changes of gas–particle flow in every computing unit to describe the properties of gas and solid phase. One does not have to monitor the trajectories of every solid particle. Therefore, Eulerian–Eulerian approach is one affordable modeling approach which could perform simulation of fluidized bed reactor at a pilot or industrial scale. For the sake of dividing the space occupied by solid and fluid, the concept of phase volume fraction (PVF) is introduced in this approach. The PVF is assumed to be continuous functions of space and time, and the sum of PVF of the gas and particle phase is equal to one [13]. This approach requires all processes at particle scale such as collision, drag, friction forces and heterogeneous chemical reactions to be solved in an Eulerian grid and included as phase interaction terms into governing equations.

Based on the Eulerian–Eulerian approach, the developed fluidezed models could include hydrodynamics, heat transfer and chemicalreactions together in multiphase systems. In order to calculate the transient nature of a gas–solid flow, the unsteady balance equations for mass, momentum, energy and species for gas and particle phases are solved. Meanwhile, different turbulent models are incorporated to compute the gas phase turbulent effect as well as reduce the computational burden. Granular stress equations based on KTGF are used to model the constitutive relations for the solid phase. When the solid phase is modeled as two or more continuous phase (different types or size), this model is called Eulerian multi-fluid/multiphase model. Obviously, the TFM is a basis of the multi-fluid model. The governing equations and reactive submodels below mentioned are generally used to develop the TFM for DPRS. The details of the model can be found in the cited literature [13,14].

3. Results and Discussion

3.1 Two-Way Coupling Simulation

In this work, method previously proposed [15–18] based on solving diffusion equations is applied to CFD-DEM simulations, where coarse graining is used to obtain solid volume fraction, particle phase velocity, and fluid-particle interaction forces. By examining the conservation requirements, the variables to solve diffusion equations for in CFD-DEM simulations are identified. The algorithm is then implemented into a CFD-DEM solver based on Ansys Fluent and ROCKY.

The geometry of the computational domain considered in this study was a fluidized bed with a rectangular base, later it enables to calibrate the model with the laboratory stand of similar shape. At the first stage it was decided to simplify the task and does not take into account the heat transfer of solid material and other features conducted with the roasting process. The current Rocky formulation calculates the flotation and additional pressure gradient forces based on pressure derivatives extracted from the CFD solutions, which are exported from ANSYS Fluent.

In the momentum equation solved by ANSYS Fluent [7], the solution of the modified pressure field represents the static pressure variation due to the flow and the hydrostatic pressure at the reference density, as given by Equation (2). The stress tensor is given by (3) were second term on the right hand side is the effect of volume dilation

$$\frac{\delta}{\delta t}(\rho\vec{v}) + \nabla(\rho\vec{v}\vec{v}) = -\nabla p' + \nabla\bar{t} + \rho\vec{g} + \vec{F} \quad (1)$$

$$p' = p - p_0 g \cdot x \quad (2)$$

$$\bar{t} = \mu[\nabla\vec{v} - \nabla\vec{v}^T] - \frac{2}{3}\nabla \quad (3)$$

In a DEM CFD two way coupling simulation, the particles are part of the fluid flow and will affect it in a two-way interaction, i.e., the particle movement is affected by the interaction with other particles and the fluid around it while the flow is also affected by the particle presence.

On the Rocky side, the fluid flow will exert forces upon the particle, including pressure gradient (including buoyancy) force, drag force, and virtual mass force (optional). On the CFD side, the reaction of the forces upon the particles will be applied over the fluid phases. The scheme used in the Rocky DEM CFD two-way coupling module allows both solvers to run in parallel, which can considerably decrease simulation time if the solvers do not compete for resources. During setup within the Rocky coupling module interface, you can control the number of processors required by each solver to avoid resource competition.

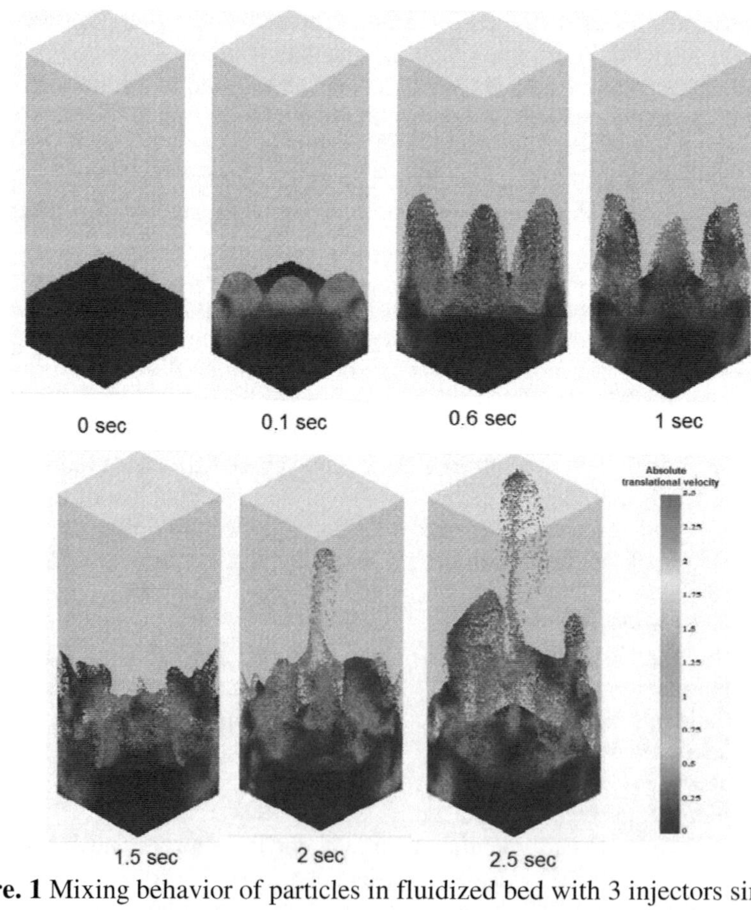

Figure. 1 Mixing behavior of particles in fluidized bed with 3 injectors simulated

The Rocky DEM-CFD two-way coupling module uses a neighbor cell averaging procedure to avoid high volume fractions in a certain cell of the CFD domain, distributing the volume of the particles that are positioned at the cell boundaries among the adjacent cells. Thus, this methodology is suitable for dealing with a large number of particles within a CFD cell and not for dealing with particles larger than the CFD cell. Therefore, it is recommended that the mesh size be larger than the largest particle size

throughout the simulation domain. Some regions with detailed geometry and strong gradients can have smaller cells but the coupling results in these regions will be less accurate.
As you can see from Fig. 1 shows the result of a computer simulation conducted using the CFD-DEM approach to investigate mixing behaviors during fluidization of solid particles (~400.000 spherical particles). The particles were differentiated by their colors in the absolute translational velocity. Particles were re-distributed rapidly and good mixing was achieved after a short period of fluidization. Such behaviors may be attributed to the presence of relative dynamic motions between particles that were induced by the flow of gas through the granular bed.

4. Conclusions

Analysis of the results shows how the various processes are closely related in the furnace. Results of the model show intensive circulation of solid and gas phase inside the furnace. Influence of solid phase at blast volume is a fairly obvious fact, as this is one of the key factors in relation to the placement of the blow injectors.
Computational studies of mixing behaviors of granular materials in gas fluidized bed systems also were conducted in this study. Methods of improving fluidized bed modeling in such systems, such as through different methods of air injection, varying the geometry of the fluidized bed, including the calculation of heat exchange between the gas and solid material, particle size distribution of various fractions, destruction and agglomeration of solids, are possible subjects of future studies.
Determination of the optimal design of the furnace, the charging device, blow nozzles, including a better understanding of the processes occurring in the furnace will allow directly to minimize the risk of excessive sintering of particles, reduce the residence time of the particles inside the machine, improve product quality, and reduce the amount of waste gases.
High temperature industrial processes is a challenge for numerical simulation as you have to take into account the multi-phase flows, interacting with each other. Recent developments within the CFD-DEM simulations allow us to consider these processes on an industrial scale in one model.
These results demonstrate how CFD model does not correspond to the actual roasting process, however this model is an important development tool for the further investigations of this type of production process. The results will provide representation for operation of existing facilities and create conditions for the design of more efficient industrial furnaces with low emissions and improved technical and economic performance.

References

[1] Koornneef J, Junginger M, Faaij A. Development of fluidized bed combustion—An overview of trends, performance and cost. Prog Energy Combust Sci 2007;33:19–55. doi:10.1016/j.pecs.2006.07.001.
[2] Samuelsberg A, Hjertager BH. An experimental and numerical study of flow patterns in a circulating fluidized bed reactor. Int J Multiph Flow 1996;22:575–91. doi:10.1016/0301-9322(95)00080-1.
[3] Crundwell FK, Moats MS, Ramachandran V, Robinson TG, Davenport WG, Crundwell FK, et al. Chapter 9 – Roasting Matte to Nickel Oxide and Metal. Extr. Metall. Nickel, Cobalt Platin. Gr. Met., 2011, p. 109–15. doi:10.1016/B978-0-08-096809-4.10009-7.
[4] Syamlal M, O'Brien TJ. Simulation of granular layer inversion in liquid fluidized beds. Int J Multiph Flow 1988;14:473–81. doi:10.1016/0301-9322(88)90023-7.
[5] A. Howley M, J. Glasser B. Hydrodynamics of a uniform liquid-fluidized bed containing a binary mixture of particles. Chem Eng Sci 2002;57:4209–26. doi:10.1016/S0009-2509(02)00361-5.
[6] Di Renzo A, Di Maio FP. Homogeneous and bubbling fluidization regimes in DEM–CFD simulations: Hydrodynamic stability of gas and liquid fluidized beds. Chem Eng Sci 2007;62:116–30. doi:10.1016/j.ces.2006.08.009.

[7] Di Renzo A, Cello F, Di Maio FP. Simulation of the layer inversion phenomenon in binary liquid--fluidized beds by DEM–CFD with a drag law for polydisperse systems. Chem Eng Sci 2011;66:2945–58. doi:10.1016/j.ces.2011.03.035.
[8] Oschmann T, Vollmari K, Kruggel-Emden H, Wirtz S. Numerical Investigation of the Mixing of Non-spherical Particles in Fluidized Beds and during Pneumatic Conveying. Procedia Eng 2015;102:976–85. doi:10.1016/j.proeng.2015.01.220.
[9] Gidaspow D. Multiphase flow and fluidization : continuum and kinetic theory descriptions. Academic Press; 1994.
[10] Briens LA, Ellis N. Hydrodynamics of three-phase fluidized bed systems examined by statistical, fractal, chaos and wavelet analysis methods. Chem Eng Sci 2005;60:6094–106. doi:10.1016/j.ces.2005.04.005.
[11] Gunn DJ. Transfer of heat or mass to particles in fixed and fluidised beds. Int J Heat Mass Transf 1978;21:467–76. doi:10.1016/0017-9310(78)90080-7.
[12] Syamlal M, Gidaspow D. Hydrodynamics of fluidization: Prediction of wall to bed heat transfer coefficients. AIChE J 1985;31:127–35. doi:10.1002/aic.690310115.
[13] Singh RI, Brink A, Hupa M. CFD modeling to study fluidized bed combustion and gasification. Appl Therm Eng 2013;52:585–614. doi:10.1016/j.applthermaleng.2012.12.017.
[14] Wang S, Guo S, Gao J, Lan X, Dong Q, Li X. Simulation of flow behavior of liquid and particles in a liquid–solid fluidized bed. Powder Technol 2012;224:365–73. doi:10.1016/j.powtec.2012.03.022.
[15] Girardi M, Radl S, Sundaresan S. Simulating wet gas–solid fluidized beds using coarse-grid CFD-DEM. Chem Eng Sci 2016;144:224–38. doi:10.1016/j.ces.2016.01.017.
[16] Traoré P, Laurentie J-C, Dascalescu L. An efficient 4 way coupling CFD–DEM model for dense gas–solid particulate flows simulations. Comput Fluids 2015;113:65–76. doi:10.1016/j.compfluid.2014.07.017.
[17] Hilton JE, Cleary PW. Comparison of non-cohesive resolved and coarse grain DEM models for gas flow through particle beds. Appl Math Model 2014;38:4197–214. doi:10.1016/j.apm.2014.02.013.
[18] Sun R, Xiao H. Diffusion-based coarse graining in hybrid continuum–discrete solvers: Theoretical formulation and a priori tests. Int J Multiph Flow 2015;77:142–57. doi:10.1016/j.ijmultiphaseflow.2015.08.014.

IOP Conf. Series: Materials Science and Engineering **221** (2017) 012018 doi:10.1088/1757-899X/221/1/012018

Wax Accumulation Prevention Method in Tubing

H N Musipov [a], **R К Akhmadulin, L N Bakanovskaya**

Industrial University of Tyumen, Russia, 625000, Tyumen, 38 Volodarskogo Str.

E-mail: [a] halim46@mail.ru

Abstract. The article describes a wax accumulation prevention method in tubing using micanite or ceramic ring heaters. A ring heater is installed on the outer surface of tubing. To supply the ring heater with electricity direct current is used. The ring heater operation mode is periodic, with the power-on time equal to the intensity of wax accumulation.

1. Introduction

The development of wax control technology and methods has had a long history. Today, as decades ago, to control wax accumulation oilmen apply the following methods: thermal, physical, chemical, mechanical and use of coatings [1-6].

Chemical methods are based on adding chemical compounds in the produced products, which reduces and sometimes completely prevents from the accumulation. Adsorption processes occurring at the boundary between the liquid phase and the surface of pipe metal are the basis of the action of wax inhibitors.

Chemical agents are divided into wetting, modifiers, depressors and dispersants [7].

Wetting agents form a hydrophilic film on the metal surface which prevents from wax crystals adhesion to the pipes; this creates a possibility of their carryover by fluid flow. These include polyacrylamide (PAA), IP-1;2;3, acidic organic phosphates, silicates of alkali metals, aqueous solutions of synthetic polymeric surface-active substances (surfactants).

Modifiers interact with wax molecules, preventing from crystal growth. This helps to keep the crystals in suspension during their movement. These properties are intrinsic to atactic propylene with a molecular weight of 2000 to 3000, low molecular weight polyisobutylene with a molecular weight of 8000-12000, aliphatic copolymers, copolymers of ethylene and compound ether with a double bond, a triple copolymer of ethylene with vinyl acetate and vinylpyrrolidone, a polymer with a molecular weight of 2500-3000.

The action mechanism of depressors is adsorption of molecules on wax crystals, which hinders their ability to aggregate and accumulate. The known depressors are "Paraflow, AzNII", alkylphenol ИПХ-, "Dorad-1A", BEO-504 TymII, "Isolyat-7" [8].

There are several best known methods dealing with asphaltene deposits (AD) actively used in oil industry. But the variety of conditions of field development and the difference in characteristics of the produced product often require an individual approach and even development of new technologies.

Dispersants – chemical agents – which provide formation of a supracolloidal system carried over by oil flow, prevent from wax accumulation on pipe walls. These include metal salts, salts of higher synthetic fatty acids, silicate-sulfanol solutions, sulfated alkali lignin [9].

Use of chemicals to prevent AD is in many cases combined with:

IOP Conf. Series: Materials Science and Engineering **221** (2017) 012018 doi:10.1088/1757-899X/221/1/012018

- the process of destruction of stable oil emulsions;
- protection of oilfield equipment from corrosion;
- protection from scaling;
- the formation of optimal gas - liquid flow structures.

A wide range of chemicals to control AD has been developed.

Along with high cost, a significant disadvantage of the chemical method is a difficulty of selecting an efficient agent because of permanent change in operating conditions of field development.

Physical methods are based on the mechanical and ultrasonic vibrations (vibration methods), as well as electric, magnetic and electromagnetic fields impact on the produced and transported products.

Vibration methods allow to create ultrasonic vibrations in the area of wax accumulation, which affect wax crystals, causing their micromovings, which prevents wax accumulation on pipe walls [8].

The magnetic field affect should be attributed to the most promising physical methods. Use of magnetic devices to prevent AD in oil production began in the fifties of the last century, but due to low efficiency did not become widespread. There was a lack of magnets for long-term and stable work under well conditions. In recent times use of magnetic fields to affect AD has become of great interest due to the occurrence of a wide range of high-energy magnets based on rare earth materials on the market. Currently, about 30 different organisations offer magnetic dewaxers [10].

It has been found out [11] that under the influence of a magnetic field in a moving fluid there occurs destruction of aggregates consisting of submicronic ferromagnetic particles of iron compounds in the concentration of 10-100 g/t in oil and associated water. Each aggregate contains a few hundreds to a few thousands of micro-particles, so the destruction of aggregates leads to a sharp (100-1000 times) increase in concentration of crystallization nuclei and salts, and to formation of micron-sized gas bubbles on the surface of the ferromagnetic particles. As a result of the aggregate destruction, the wax crystals precipitate in the form of fine, voluminous, stable suspension; and the accumulation growth rate reduction is proportional to the mean size reduction of the wax crystals precipitated as solids along with resins and asphaltenes. The formation of gas microbubbles in crystallization nuclei after magnetic treatment provides, according to some researchers, gas-lift effect, leading to some increase in well production.

Currently they use hot oil or water as thermal fluids; superheated steam; surface and borehole electric furnaces; electric dewaxers (induction heaters) operating inside boreholes, agents leading to exothermic reactions when interacting [12, 13].

[1]There are innovative wax control technologies [14].

These methods, with slight variations, and sometimes without them, are used in wells with various types of operation.

2. Research

Thermal methods are based on the ability of wax to melt at temperatures above 50°C and flow down from the heated surface.

Currently there are applied the technologies which use:

- hot oil or water as thermal fluids;
- superheated steam;
- surface and borehole electric furnaces;
- electric dewaxers (induction heaters) operating inside boreholes;
- agents leading to exothermic reactions when interacting.

The thermal fluid technology provides heating fluids in special heaters (portable boiler) and delivering them to the borehole by means of normal or reverse circulation. Reverse circulation is more preferable because it prevents formation of wax plugs that often occur during normal circulation.

The disadvantages of these methods are their high energy output, electricity and fire risk, unreliability and low efficiency.

To create the required temperature a special heat source is required, which can be placed directly into an accumulation zone; or a heat-containig agent is to be developed at the wellhead.

IOP Conf. Series: Materials Science and Engineering **221** (2017) 012018 doi:10.1088/1757-899X/221/1/012018

These methods have certain disadvantages. Thus, the use of hot oil and water or superheated steam leads to heating the tubing section from the wellhead to the wax accumulation zone, in other word, to thermal fluids energy losses. To provide the temperature of over 50°C in the wax accumulation zone thermal fluids have to be heated at the surface to the temperature 3-4 times exceeding the required one to melt wax. This method requires stopping the well. This method is expensive and energy consuming with low efficiency

A more efficient method is creating the required temperature in the very wax accumulaytion zone. For this purpose a special heat source that can be placed directly in this zone is required

Induction heaters are used as such sources. The induction heaters design involves the use of high frequency currents (HFC) to heat the tubing wall. This is achieved by using an inductor installed in the wax accumulation zone. A generator is installed at the wellhead. The method is not electrically safe, and requires careful HFC source insulation from the well metal structures.

The authors [15] suggest a method of creating the required temperature in the very wax accumulation zone using ceramic or micanite in-line ring heaters.

Ceramic or micanite in-line ring heaters (Fig. 1) are suggested to be installed on the tubing outer surface at the wax accumulation zone. Ring heaters are ideal for heating tubes and cylinders. Ceramic ring heaters have a high specific capacity for heating, so they can be used at higher temperatures.

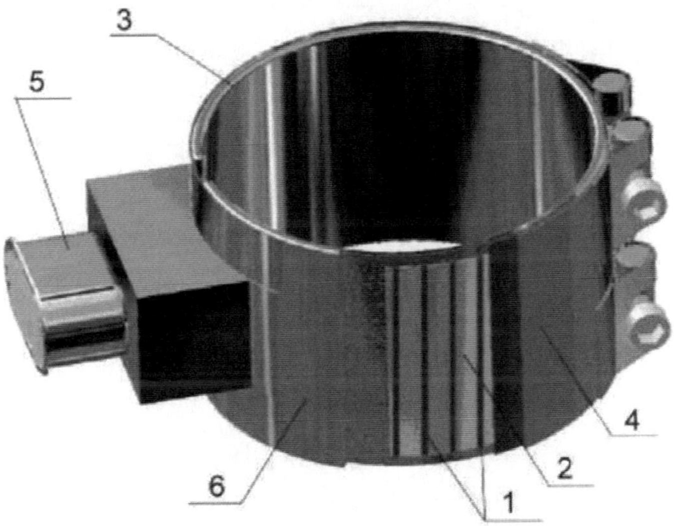

Рис. 1. Design of a ring heater
1 – resistance tape; 2 –magnetic core; 3 - stainless steel case; 4 –micanite layer; 5 - electrical input; 6 - clamp

The rings rim diameter is equal to the tubing external diameter; the width should be equal to the interval of wax accumulation on the inner walls of the tubing. If the interval of wax accumulation exceeds the width of the ring resonator, the latter is made compound.

Permissible specific load of the ring heater with micanite insulation can reach 4.5 W/cm^2, and with ceramic insulation – up to 8 W/cm^2.

The operating temperature generated by the ring heater with micanite insulation can reach 350°C, and with ceramic insulation - 600°C.

The production technology allows to reduce energy consumption by 30% and greatly increase the heater useful lifetime.

The heat transfer from the heating element to the tubing heated surface is performed by the contact method.

Wire or high-resistance alloy tape are used as the heating element in the ring heater.

The insulation materials are micanite or ceramics. In heaters with micanite insulation the resistance tape (Fig.1, 1) is wrapped around the core (2) of the required shape and placed in a stainless steel case (3). A layer of micanite (4) is laid between the core and the case, which provides complete electrical insulation of the heater. Then, the heater is shaped into a cylinder, and the electrical input is installed (5). The reliable contact of the heater with a heated tubing surface is provided with a clamp (6) with tightening screws.

Ceramic ring heaters are similar to micanite ones, but are applied at higher loads and temperatures up to 700°C (Fig. 2).

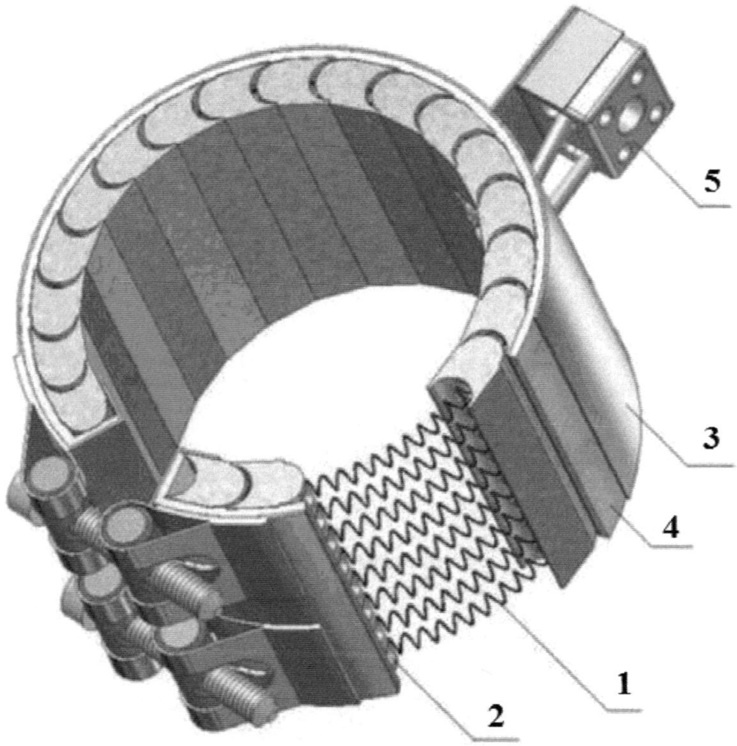

Рис. 2. Energy-conserving ring clamp heater
1 – spiral structure of X20H80 wire; 2 – ceramic insulators;
3 – tighting clamp; 4 – heat-insulating layer; 5 – terminal lead

3. Conclusion

Energy-conserving ring clamp heaters save up to 20% of electricity. When installing and operating ring clamp heaters it is required to provide maximum contact of the heating part of the heater and the heated surface for long and reliable service. For this purpose high temperature mounting paste is often used.

Heating control is possible due to a program, providing a periodic electricity supply through a controller.

References

[1] Khaibullina K 2016 Technology to remove asphaltene, resin and paraffin deposits in wells using organic solvents *SPE Annual Technical Conference and Exhibition ATCE 2016 (*Dubai) Code 12406

[2] Zlobin A 2017 The mechanism of magnetic activation of oil for well protection against asphaltene deposits *J. Neftyanoe Khozyaystvo - Oil Industry* **1** 52-56

[3] Pate B and Stanley R 2010 Challenging global capillary applications for enhanced production *Society of Petroleum Engineers - IADC/SPE Asia Pacific Drilling Technology Conf. 2010 - Value, Resources and Innovation: Sustaining the Next Decade* (Ho Chi Minh City) pp 289-299

[4] Khisamov R, Khuzin R, Andreev V, Dubinsky G and Miyassarov A 2015 Prospects of increase in efficiency of development of reservoirs containing high-viscosity oil with the use of energy-saving and resource-saving technologies *J. Neftyanoe khozyaystvo - Oil Industry* **4** 52-55

[5] Murtaza S, Al-Ruwaily A, Taqi A, Qahtani S, Chacon A, Gonzalez A and Bustamante E 2013 Wellbore asphaltene cleanout using a new solvent formulation in a horizontal open hole oil producer in carbonate reservoir of north Ghawar field-scripting a success story *18th Middle East Oil and Gas Show and Conf. 2013: Transforming the Energy Future MEOS 2013* vol 3 (Manama; Bahrain) pp 2059-70

[6] Musipov H, Yerka B, Illarionova E and Ivanova Y 2009 Providing environmental safety of the oil-gas enterprises *J. Neftyanoe khozyaystvo - Oil Industry* **1** 92-94

[7] Golonskiy P 1960 Wax control in oil production (Moscow: Gostoptekhizdat) p 88

[8] Persiyantsev M 2000 Oil production under abnormal conditions (Moscow: LLC "Nedra-Business Centre") p 653

[9] Ibragimov G 1986 Chemicals for oil production: Ref. book for a worker, ed G Ibragimov, V Sorokin and N Khisamutdinov (Moscow: Nedra) p 240

[10] Kovach V Alivanov V and Shaidakov V 2002 Magnetic activation of a fluid as a method of corrosion protection *Neftyanoe khozyaystvo - Oil Industry* **10** 126-128

[11] Lesin V 2001 New generation magnetic dewaxers *J. Izobreteniya i Ratspredlozheniya v Neftegazovoy Promyshlennosti* **1** 18-20

[12] Tiwari S, Verma S K, Karthik R, Singh A K, Kumar S, Singh M and Kothiyal M D 2014 In-situ heat generation for near wellbore asphaltene and wax remediation *Society of Petroleum Engineers - International Petroleum Technology Conf. 2014, IPTC 2014: Unlocking Energy Through Innovation, Technology and Capability, IPTC 2014 vol 5* (Doha; Qatar) pp 3981-89

[13] Pacheco-Sanchez Juan H and Mansoori G Ali. 1998 In situ remediation of heavy organic deposits using aromatic solvents *68th Annual Western Regional Meeting* (Bakersfield, CA, USA) p 10

[14] Viktorovich K O and Alekseevich V D 2014 Advanced stimulation technologies for enhanced unconvencional oil and gas recovery *Society of Petroleum Engineers - SPE Russian Oil and Gas Exploration and Production Technical Conf. and Exhibition 2014, RO and G 2014 - Sustaining and Optimising Production: Challenging the Limits with Technology vol 2* (Moscow) pp 1748-55

[15] Musipov H 2012 Patent No 22450117 RF C1. A method of heating gas-liquid mixture in a well to prevent wax accumulation in tubing ed H Musipov and Y Savinykh bul. 13 (Claimed 01.10.2010 Published 10.05.2012)

IOP Conf. Series: Materials Science and Engineering **221** (2017) 012019 doi:10.1088/1757-899X/221/1/012019

Increasing the Adaptive Capacity of the Organism When Exposed to Adverse Environmental Factors Through Phytoadaptogens

N V Babii[1], V A Pomozova[2], T F Kiseleva[2], V O Romanenko[3, a]

[1] Amur State University,
Russian Federation, Blagoveschensk, Ignat'evskoe shosse, 21
[2] Kemerovo technological Institute of food industry (University),
Russian Federation, Kemerovo, b-r Stroiteley 47
[3] Yurga technological Institute branch of Tomsk Polytechnic University,
Russian Federation, Yurga, 26 Leningradskaya street.

E-mail: [a]romvas-1989@mail.ru

Abstract. The Problem of improving the quality of life of the population in modern conditions is the most relevant. The level of human health largely depends on the quality of the environment. Of special importance are the risk factors for the spread of diseases like environmental pollution, social conditions and bad habits. To the emergence of ecologically dependent diseases leads the combined impact of technogenic, social-economic, natural-climatic factors. The introduction of the principles of biotechnology, nanotechnology and innovations in industrial processing of plant resources, especially natural adaptogens, contributes to the production of balanced, ecologically clean food products functional purpose, the use of which will significantly minimize the adverse effects of the environment on humans.

1. Introduction

Today, human activity gets out of geo-ecological processes leads to a change in the natural biogeochemical cycles on earth, the violation of the ecological balance in the biosphere that in turn affects the man himself. The food chain get alien non-food components, which can lead to the development of a number of ecologically dependent diseases: allergies, malformations, mutational changes, intoxications, tumors, etc.

The last led in the present, the shift from environmental protection to protection of human health as the main resource and engine of progress. The level of human health largely depends on the quality of its habitat. Environmental factors 18-20% determine the health status and are in second place after lifestyle.

The lack of effectiveness of existing means for the prevention of ecologically dependent diseases, necessitated the creation of new types of functional products. Development and production of food products with functional properties, is a global trend. The production of functional products allows efficient use of nutrients as a source of energy and plastic material for various organs and systems to maintain the native biologically active substances, to reduce the content of substances which have a toxic effect on the body.

Throughout the history of mankind wild plant raw materials used for cooking a variety of preventive tools that can stimulate the body's defenses and thereby enhance its performance as a result of their actions are activated functions of organs and systems [1]. This effect can be achieved with the help of stimulant drugs – adaptogens. The action of these substances on the processes of cellular metabolism is the synthesis of structural substances and of energy-rich compounds. The result is a physiological mobilization of protective forces. These substances, expanding bottlenecks metabolism, prevent the violation of the energy and plastic processes in the tissues. They are capable of long time to maintain under extreme conditions, the constancy of the internal environment that, according to conventional ideas is a prerequisite for healthy existence of the body. The ability of substances to activate the body's defenses and thereby enhance their resistance to extreme agents gave the reason to allocate them in separate group – adaptogens.

The development of the production of functional foods based on natural adaptogens - the most important scientific direction, based on the current understanding of the physiology of adaptation, homeostasis, regulations for biological Cybernetics and the theory of functional food. The most significant trend in these studies is the development of approaches, criteria and methods to assess adaptive capacity of the organism.

2. Review and Analytics

Natural adaptogens enhance mental performance, improve psychomotor activity, humoral immunity.

When using drugs ginseng (lat. Panax zingiberensis) and Siberian ginseng (lat. Eleutherocóccus), the reduction of the expenditure of muscle glycogen, lactate content increased to a lesser extent. The positive effect of Rhodiola rosea (lat. Rhodíola rósea) and Eleutherococcus senticosus (lat. Eleutherocóccus senticósus) in exhausting muscular exertion associated with their stimulating effect on the plastic exchange, resulting in the increase in muscle RNA. High efficiency has schisandra chinensis (lat. Schisándra chinénsis) increase the body's resistance to hypothermia, physical and mental stress, and various types of rose hips, which have a beneficial effect on the human body [2, 3].

A kind of plant adaptogens, the so – called biostimulants, which include the extract from the aloe leaf (lat. Álоё), juice of stems of Kalanchoe (lat. Kalanchoë), etc.

According to some researchers [5, 6], the pharmacological effect due to the presence of saturated and unsaturated aliphatic dicarboxylic acids, aromatic phenolic acids and other natural compounds.

Herbal adaptogens reduce the severity of the lesions with ionizing radiation and emotional stress, applied in the complex treatment of debilitated patients who have reduced body resistance, in cancer patients, to enhance physical stamina in athletes and mental health, in occupational exposure, during work in hot shops, in conditions of lack of oxygen in the atmosphere, and at high concentrations of toxic substances in the air of working zone [5].

Adaptogens have a mild but steady antihypoxic and antioxidant action.

It is proven that stress can contribute the development of hypertension in genetically determined or caused by other factors, the inadequacy of mechanisms of regulation of water-salt metabolism and vascular tone. Based on the study of the literature found that 2/3 of people cancer diseases in the next 2-4 years before the disease experienced severe stress caused by prolonged depression [7].

Adaptogens can cause and support the body's need adaptive response to increase resistance in each case. Such versatility is determined by the ability to adjust for stress reactions [8]. In the absence of stress on the background of a relatively normal condition of the body was also highlighted certain effects of adaptogens expressed to a lesser degree. However, the value of adaptogens on the body is extremely large, since we are talking about creating in the body with adaptogens a "margin of safety" - provision of health, which is crucial for disease prevention [9].

In this respect, of particular value represent multicomponent mixture of vegetable raw materials, in which consider three rules: a tonic effect, stimulating effect and influence on the particular disease,

IOP Conf. Series: Materials Science and Engineering **221** (2017) 012019 doi:10.1088/1757-899X/221/1/012019

which can be regarded as one of the major factors contributing to the increase of nonspecific resistance of the organism.

The aim of this work is the development of principles of creation of new kinds of functional drinks based on adaptogens from plant material grown in the territory of the Russian Federation and the practical approbation of the proposed methods.

3. Results

A special role in the practice of healthy eating is given to drink, considering their popularity among the population, an enrichment of biologically active substances for various applications.

Certain priorities in the production of enriched drinks is the use of medicinal vegetative raw materials, containing in its composition a wide range of substances pharmacological selectivity [9].

To implement the process of development and implementation of enriched product can only be based on clearly defined, scientifically based and proven medical and biological principles that determine the most significant issues of the development, production and sales, vitamin-enriched foods.

When designing one of the important tasks of the food systems is ensuring the optimal set and the ratio RX of the components for the development of new functional foods, including beverages for various categories of the population.

In the present work supplemented by, principles of design of functional beverages, the methodology for the production of healthy food developed N. N. Corneum, E. P. Viktorova, O. V. Evdokimova [10].

The application of the proposed principles in practice allows to exclude empirical approach in the search for raw materials and determine various aspects of feasibility of the combination by calculating the participation criteria of the individual components of the formulation in the formation of quality new products.

This principle makes it possible to rank the product components according to their contribution to the properties of the whole structure, to propose both analytical and synthetic conclusion, to assess their relationship.

The main design principles of the drinks based on natural adaptogens is presented in figure 1.

The principle functionality is a fundamental attribute of the system. Any system interacts with the environment, i.e. has a certain set of functions. This principle is based on the therapeutic orientation of certain properties of the drink as a prophylactic measure.

The principle of compatibility takes into account the pharmacological compatibility of recipe components do not cause undesirable interactions, availability of substances-synergists to achieve private goals, such as persistence, indicators of quality, including sensory characteristics, the optimum composition of the product, absorption.

The principle of minimization. The prosperity of the human body depends on a complex of factors, each of which organism there is a certain range of tolerance. Tolerance occurs when repeated administration of a given amount of a substance causes a reduced effect or when to obtain the effect previously achieved smaller dose requires a consistent increase in the number of administered substances.

Thus, this principle is considering the possibility of receipt of biologically active substances in the human body with the product (including a drink) in a specific quantitative range below the physiological minimum, but not above the recommended standards and is based on the recommended levels of food consumption and biologically active substances.

The principle of the combinatorial hierarchy the entity is in the set of operations aimed at physico-chemical, colloidal, technological and other changes in the composition and structure of the product with the aim of obtaining the specified performance.

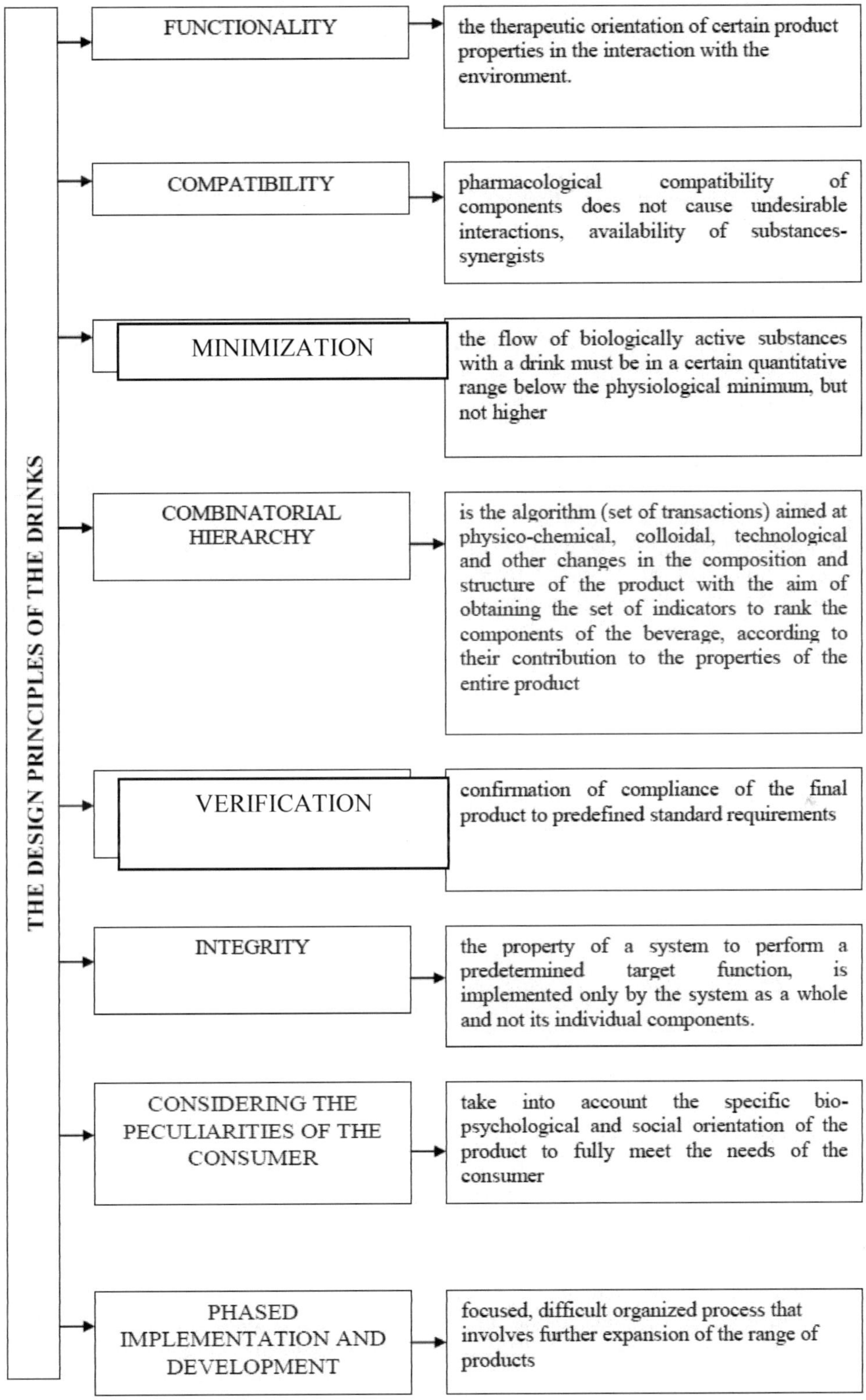

Figure 1 – Basic design principles based drinks natural adaptogens

IOP Conf. Series: Materials Science and Engineering **221** (2017) 012019 doi:10.1088/1757-899X/221/1/012019

The principle of verification be to ensure that the confirmation of compliance of the final product must conform to predefined standard requirements.

The principle of integrity involves the separation of the object of study of holistic education, i.e., distinguish it from other phenomena, from the environment. This can only be done through the identification and evaluation of the characteristics of phenomena and comparison of these properties with the properties of its elements.

The principle of the personal characteristics is the need to identify and address the BIO-psychological and social orientation of the product to fully meet the needs of the consumer (age, gender, working conditions, and price preferences of the user, the purpose of the application of the product).

The principle of gradualism and development is a purposeful, difficult organized process in which there are different stages. Each of them has its objectives, tasks and methods of correction. Consistently formed the preconditions for expanding the range of products (including by and drinks). The range of products will cover all segments of the population and to ensure the effectiveness of the prevention of various diseases.

4. Practical approbation of the developed methods

One of the options for obtaining functional beverages can be fermented beverages obtained as a result of unfinished alcohol and/or lactic acid fermentation of kvass wort concentrate or fruit and berry musts. Thus, in the development of technologies new types of functional beverages, the starting point was the technology popular in Russian Federation bread fermented beverage – kvass.

As an adaptogenic basis for the preparation of functional beverages used an extract of schisandra chinensis (lat. Schisándra chinénsis), Siberian ginseng (lat. Eleutherocóccus), concentrated juices of black currants (lat. Ríbes nígrum) and Apple juice (lat. Mālus).

The result of extraction of biological components go into solution and become more accessible and digestible. But, as vegetable raw material (Eleutherococcus) is significantly different from the schisandra fruit, it is considered appropriate to investigate the extraction process separately for each type of raw materials used. The first stage of the research was the selection of the extractant. Any extractant was selective, chemically and pharmacologically (from the point of view of security) to be weak, to be available and cheap.

Infusion of Eleutherococcus carried out with water for 30-120 min at ratio raw material:extractant 1:50-1:100. In the obtained extracts was determined biologically active substances to the main component - eleuterozida B (table 1).

Table 1 – Influence of parameters of the extract of Eleutherococcus on eleuterozida B

The conditions of extraction	The contents of eleutheroside B in % when the ratio of raw materials-extractant	
	1:50	1:100
120 min	0.061	0.063
60 min	0.057	0.059
30 min	0.059	0.057
4-fold extraction for 30 min.	0.064	0.064

Comparison by Student's t test and Fischer (P=0.95) showed no significant influence of selected parameters of the methodology on the results of the analysis (F_H. < Fkr. and $t_H < t_{kr}$). In light of this methodology was chosen extraction time of 30 minutes. The optimal ratio of raw materials-extractant is 1:50.

Extract of Siberian ginseng increases endurance, performance and participates in toning the body, but also positively affects the mental and physical activities.

Extracts of schisandra chinensis were prepared, taking into account the coefficient of water absorption, which was determined experimentally. For the preparation of infusions used chopped dried raw material passing through a sieve with a hole diameter of 2 mm. the Experiment was conducted on 4 samples subjected to one form of drying: convection, infrared. In a series of pilot experiments were the optimal parameters of extraction: duration of 6 h, temperature 35 0C, the extractant 1:10 with occasional stirring of the mixture at intervals of 30 minutes. The duration of extraction for six hours due not only to output the maximum number of biologically active components, but also the "ripening" of the extract, expressed in the formation of the fullness and harmony of taste. The main components of the extract having biological activity are polyphenolic nature, in this regard, the emphasis in the analysis of biologically active substances (BAS) were made on these compounds [10]. The content of biologically active substances in the water extract from dried fruits of schisandra chinensis obtained by convective drying (table. 2).

Table 2 – Content of biologically active substances in the extract from the fruit of schisandra chinensis

The method of drying	The content of biologically active substances, %			
	tannins	phenolic compounds	anthocyanins	organic acids
Natural	0.021	0.008	0.0015	1.26
Convective	0.085	0.025	0.0072	2.18

As can be seen from the above data, the content of biologically active substances in the extracts obtained from fruits dried by convective method, 2 - 4 times higher than naturally dried.

The obtained data confirm that the selected extraction parameters are optimal, because they retain the maximum number of biologically active substances (80-93 %) for all analyzed indicators.

The obtained extracts have a high nutritional value and can be used as sources of biologically active substances upon receipt of the drinks of a functional purpose.

The authors developed a recipe and technology of kvass functional purpose, allowing their use with minimal time to adapt to adverse environmental factors.

Qualitative and quantitative composition of the components kvass was determined by sensory characteristics, and taking into account recommended standards of the contents of biologically active substances in the finished product.

The quantitative composition of the components kvass is given in table 3. To increase the vitamin value in the sample No 1 was added concentrated black currant juice, and in sample No 2 – concentrated Apple juice.

Table 3 – Consumption of the components per 100 dm^3

Name of raw materials	Sample 1	Sample 2
Kvass wort concentrate, kg	28.6	28.6
Sugar, kg	50	50
Yeast, kg	0.15	0.15
Extract of schisandra chinensis, dm^3	1.2	-
Siberian ginseng extract, dm^3	-	1.2
Concentrated black currant juice, dm^3	0.3	-
Juice concentrate Apple juice, dm^3	-	0.3

Fermentation was carried out in different ways: using only the yeast Fermipan soft; yeast Fermipan soft together with dressing Ultravit; lactic acid bacteria (bulgaricus) together with a yeast feeding Ultravit and mixed ferment with Baker's yeast Fermipan soft (0.025 %) and lactic acid bacteria (Bulgarian Bacillus 0.015 %). Dosage of insertion of yeast and lactic acid bacteria were established on the basis of recommendations for their use in the production of kvass.

The fermentation was conducted at 30 °C to a solids content of 5 %. The dynamics of the digestion process are shown in figures 2 and 3.

From the data presented in figures 2 and 3 shows that the fermentation process proceeds most intensively during the fermentation the yeast mixture and nitrogen top-dressing "Ultrafit", as well as yeast feeding and lactic acid bacteria (48 hours).

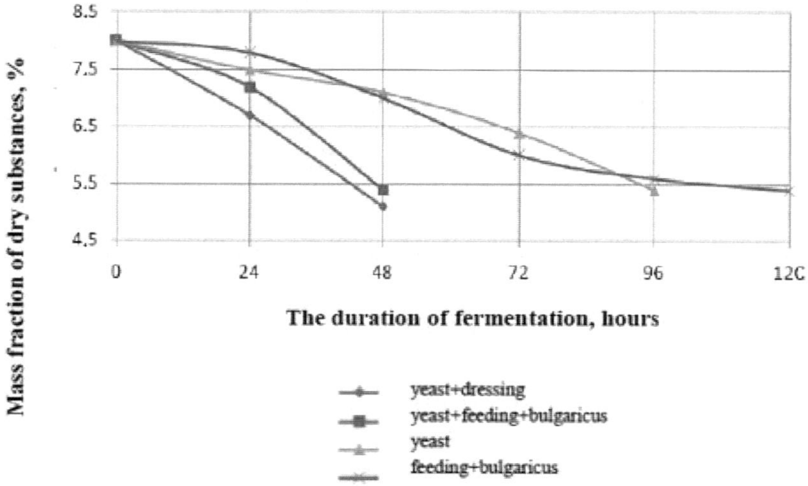

Figure 2 – Dynamics of digestion model No 1

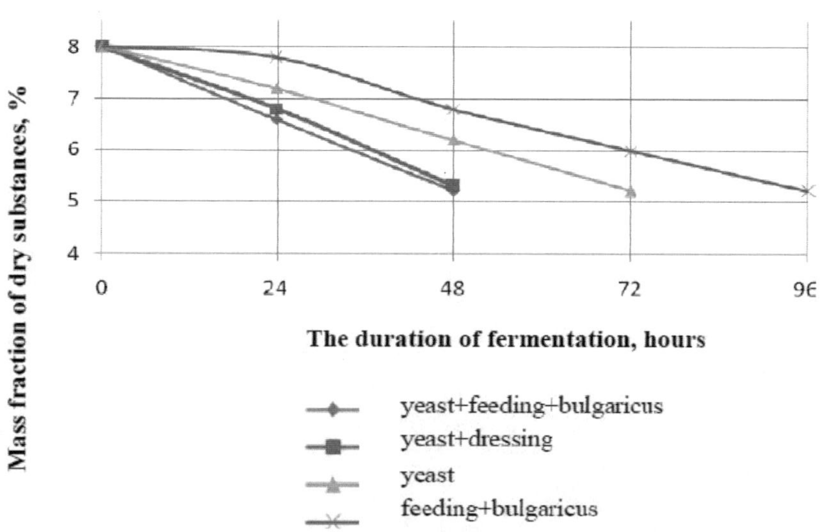

Figure 3 – the Dynamics of the digestion model No 2

This can be explained by the fact that when making nitrogen fertilization environment is enriched with nitrogen, necessary for vital activity of yeast. In the second case, the must is enriched not only with nitrogen but also lactic acid, which is lactic acid bacteria during the fermentation of sugars and also creates favorable conditions for vital activity of yeast.

Physical and chemical indicators of the quality of the resulting beverage are shown in table 4.

Table 4 – Physico-chemical indicators of quality beverages with adaptogenic properties

Name of the indicator	Sample 1	Sample 2
Mass fraction of dry substances, %	5.2±0.1	5.3±0.1
Mass fraction of alcohol, %, no more	1.05±0.02	0.98±0.02
Mass fraction CO_2, %, not less than	0.4±0.1	0.4±0.1
Acidity, K. ed.	4.1±0.2	3.7±0.2
The contents of polyphenolic compounds, mg/100 cm^3	51.2±6.0	29.3±6.0
Mass fraction of vitamin C, mg/100 cm^3	29.2±1.2	25.4±1.2

5. Conclusion

1. Added methodology for the development of functional foods, the design principles of the drinks based on herbal adaptogens.

2. Developed formulation and technology of production of beverages that do not contain synthetic additives. As dyes were natural pigments included in the composition of plants. The parameters of the most rational parameters of digestion.

Vitamin C content in the obtained fermented beverages allows you to provide up to 50 % of the daily requirement of the human body ascorbic acid (daily requirement of 90 mg), with one-time consumption of 250 cm^3 of the drink. In addition, the drinks are a good source of intake of polyphenolic compounds, which represents added value for consumers. Compounds such as leucoanthocyanins, rutin and a number of close connections belong to the P-active and possess biologically active beginning.

Polyphenolic substances, as natural synergists, ascorbic acid, increase the strength of capillaries and reduce their permeability. Anthocyanins are able to prevent or reduce the negative consequences of radiation injuries. Regular supply the human body sufficient amounts of polyphenolic substances can be one of the measures of prevention imperceptible, but continuous exposure to radiation.

References

[1] Kirke, D. A.; Smyth, T. J.; Rai, D. K. The chemical and antioxidant stability of isolated low molecular weight phlorotannins, Food Chemistry 221, P.1104-1112.

[2] Das, Quail; Islam, Md Rashedul; Marcone, Massimo F. Keith Warrinera, Moussa S. Diarra. Potential of berry extracts to control foodborne pathogens, Food Montrol Vol. 73, P. 650-662.

[3] Laaksonen, Oskar; Knaapila, Antti; Niva, Tuija. Sensory properties and consumer characteristics contributing to liking of berries, Food quality and preference Vol. 53, P. 117-126.

[4] Cristina Calejaa, Lillian Barrosa, Amilcar L. Antonioa, M. Beatriz P.P. Oliveirac, Isabel C.F.R. Ferreira, A comparative study between natural and synthetic antioxidants: Evaluation of their performance after incorporation into biscuits, Food Chemistry Vol. 216, P. 342-346.

[5] MacLeod, Alan; Jones, Glyn D.; Anderson, Helen M. Plant health and food security, linking science, economics, policy and industry: OECD Meeting: 2014 Food security: 1 P. 17-25.

[6] Erenler, Ramazan; Sen, Ozkan; Aksit, Huseyin. Isolation and identification of chemical constituents from Origanum majorana and investigation of antiproliferative and antioxidant activities, Journal of the science of food and agriculture Vol. 96, P. 822-836.

[7] Deng, Ke-jun; Zang, Li-li; Lan, Xiu-hua. Antioxidant Components from Cape Gooseberry, Journal of food processing and preservation Vol. 40, 5 P.893-898.

[8] Sergeeva, I.Yu. Technological supplementary means for improvement of beverage technology // Foods and raw materials. Vol. 3, 1, 2015. P. 41-49.

[9] Kiselev V.M., Kiseleva T.F., Terentyev V.A. Consumer demand descriptors for the emerging wine market // Transformations in business and economics. 2016. vol. 1(15). P. 160-172.

[10] N.N. Kornen, E.P. Viktorova, O.V. Evdokimova , Methodological approaches to the creation of healthy food / Voprosy Pitaniia. 2015. Vol. 84, 1. P. 95-99.

IOP Conf. Series: Materials Science and Engineering **221** (2017) 012020 doi:10.1088/1757-899X/221/1/012020

Innovative Equipment and Production Method for Mixed Fodder in the Conditions of Agricultural Enterprises

U K Sabiev[a], **E V Demchuk, V V Myalo, A S Soyunov**

Omsky State Agrarian University, 644008, Omsk, Fizkulturnaya street 1

E-mail: [a] uk.sabiev@omgau.org

Abstract. It is recommended to feed the cattle and poultry with grain fodder in the form of feed mixture balanced according to the content. Feeding of grain fodder in the form of stock feed is inefficient and economically unreasonable. The article is devoted to actual problem – the preparation of mixed fodder in the conditions of agricultural enterprises. Review and critical analyses of mixed fodder assemblies and aggregates are given. Structural and technical schemes of small-size mixed fodder aggregate with intensified attachments of vibrating and percussive action for preparation of bulk feed mixture in the conditions of agricultural enterprises were developed. The mixed fodder aggregate for its preparation in the places of direct consumption from own grain fodder production and purchased protein and vitamin supplements is also suggested. Mixed fodder aggregate allows to get prepared mixed fodder of high uniformity at low cost of energy and price of production that is becoming profitable for livestock breeding. Model line-up of suggested mixed fodder aggregate with different productivity both for small and big agricultural enterprises is considered.

Introduction

The national project of Russia in the field of agriculture provides for intensive development of livestock breeding. In common production process livestock breeding spends more than half of cost for mixed fodder.

It is proved by scientific researches and practice that increasing productivity of cattle and poultry depends much upon the quality of mixed fodder. Feeding of grain fodder in the form of feed mixture is inefficient and economically unreasonable. Simple feed mixture from several kinds of grain fodder, balanced according to the content gives considerably greater effect than simple stock feed prepared from one crop [1].

Complete mixed fodder, balanced according to the main feed elements, microelements and vitamins are in 25-39% more effective than the cereal feeds [2]. Feed grain can be processed in mixed feed, developing home-grown mixed fodder production. It allows to reduce cost for purchases of raw material, its transportation, to use grain fodder and purchased expensive protein and vitamin mineral components more rationally and to provide continuously collective, peasant (farm) holdings with home-grown mixed fodder. That is why the production of mixed fodder in the places of direct consumption is becoming profitable for livestock breeding.

Objects and methods

At present many agricultural enterprises of different forms of property abandon from mixed fodder purchase on a big feed plants because of high prices and transportation expenses. That is why

IOP Conf. Series: Materials Science and Engineering **221** (2017) 012020 doi:10.1088/1757-899X/221/1/012020

one of the solution variants of this task in these conditions are to organize mixed fodder production on the base home-grown grain raw material and purchased protein and vitamin supplements and premixes

During the preparation of mixed fodder in the conditions of agricultural enterprises usually use equipment of production units of EMFPU type (equipment of mixed fodder production units) or EPUMFP (equipment of production units for mixed feed production). Such production units for mixed fodder occupy big areas, require high investments and have high energy and metal consumption but do not provide mixed fodder of high quality.

In connection with the above, many joint stock companies of our country need in rather cheap and simple in construction and maintenance arrangements for mixed fodder production from home-grown feed raw material. This problem is particular sensitive in connection with the development of peasant and farm holding, where reliable and available in price technique for mixed fodder production are required [3].

There are many type of equipments for such conditions of mixed fodder production. Although it is difficult for commodity producers to see into effectiveness and applicability of usage these equipments. Well-known and stock-produced mixed fodder aggregates and arrangements (UMK-F-2, KA-4, mixed fodder production units of small size «Uralets», «ITAI» and others) have series of disadvantages [4, 5, 6, 7, 8]. In these aggregates – finished operation of mixing is done by augers. As you know, during the operation both horizontal and vertical auger rotates inside the cylindrical section, where there are some quantity of mixture (mixture of different components). It is necessary to apply considerable effort for auger rotation during the mixing. It requires much energy but it doesn't often provide mixed fodder of high quality.

It is obvious that it is necessary new, innovative technologies and processes for mixed fodder preparation, providing the growth of effectiveness of feed preparing equipment.

The review of modern machines condition of vibrating action in different branches of economy shows the advantages of vibromethod from the standpoint of reduction of efforts and energy consumption, increasing the productivity of equipment and quality fulfillment of technological processes. Vibrating processes and machines are considered as technology of the future [10, 11]. All these advantages open them wide perspectives for creation new mixed fodder technological processes.

Vast researches on development of vibrating and percussive - centrifugal feed preparing machines are carried out in the Southern-Ural agrarian university headed by professor N.S. Sergeev [8]. The vibropercussive technology of mixed fodder production is carried out by professor I.Ya. Fedorenko in the Altaj SAU [6, 7]. Researches in this field showed practical reason and economic effectiveness of usage of vibrating and percussive attachments for processing of feed materials.

All foregoing allows to make the conclusion that it is necessary to involve vibrating processes and machines in technology of preparing bulk feed mixture. Also at present time the effective technologies and technical equipments, intensified the process of preparing concentrated feeds for cattle feeding in the conditions of agricultural enterprises are absent.

Specific of mixed fodder preparation technology in the conditions of agricultural enterprises allows to examine formalized relations of process holding in working equipment (for separation, chopping, dosing and mixing) with an account of vibrating and percussive influence on the bulk feed (components of mixed fodder) and their physical- mechanical properties as well as influence of these conformities to law on qualitative index of prepared mixed fodder.

Mixed fodder arrangement of small - size on the base of serial mixed fodder aggregate UMK-F-2 deserves great attention [3, 9]. Feature of suggested construction of mixed fodder arrangement of small - size is that it is provided with vibrating mixer with step mixing elements in the form of semi-spheres, set after crush chamber. The diameters of mixing elements of vibrating mixer are increased from the foundation to the edge of teeth according to the arithmetic progression law and established according to chess order. Increasing the homogeneity degree of mixture during simultaneous reduction of energy consumption and duration of mixed fodder preparation process are provided due to it.

IOP Conf. Series: Materials Science and Engineering **221** (2017) 012020 doi:10.1088/1757-899X/221/1/012020

Investigation results

Technology of mixed fodder preparation consists of several operations. Expenses of energy consumption take the chopping and mixing processes. They became important in technological lines on processing of plant raw material. Another operations (for example separation and dosing) are auxiliary. The main functions provide efficiency of all object but auxiliary ones characterize the means of achievements of the main functions and promote their realization.

According to the rates of standards and normative-technical documents, used in mixed fodder preparation in the conditions of agricultural enterprises, technical solutions must provide some requirements: increasing the productivity and division of grain lots; increasing the uniformity of granulometric grinding product content; increasing the effectiveness of dosing and mixing of fodder components. That is why criterions attracting qualitative and quantitative indexes of preparation: increasing the quality and reduction of energy consumption of mixed fodder preparation process are chosen for estimation of effective functioning of technical equipments for mixed fodder preparation.

To achieve high quality of technical equipment operation due to intensification of mixed fodder preparation processes they must provide the maintenance of all requirements. Such functions as external impact, conditions of mixed fodder preparation equipment and technical impact on quality are influenced during the operation of each mixed fodder preparation aggregate.

Technological process of mixed fodder preparation in the conditions of agricultural enterprises is complex multiparametrical system where the scheme of functioning has hierarchical structure, including subsystem presented the models of some processes (separation, chopping, dosing and mixing), phenomenon and their relationship.

Suggested mixed fodder preparation processes may be considered as multidimensional dynamic system with many input and output parameters, which operates in the conditions of continuously changed external influences. Each type of operation in suggested technology of mixed fodder preparation is subsystem and carried out in specific sequence. The result of their work is increasing the quality and reduction of energy consumption of mixed fodder preparation process.

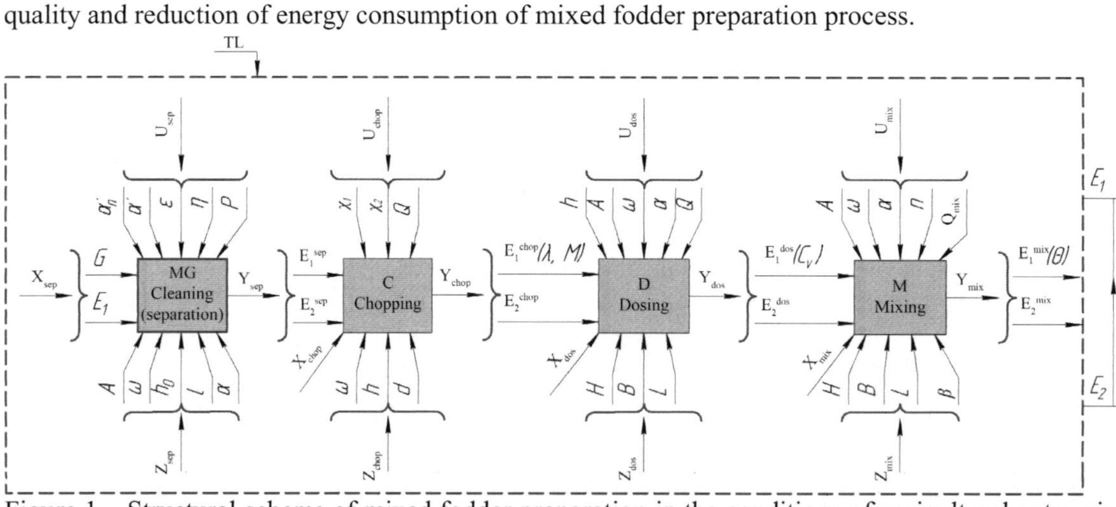

Figure 1 – Structural scheme of mixed fodder preparation in the conditions of agricultural enterprises: GM – grain cleaning machine; C – chopper, D – vibrodoser, M – vibromixer; TL – technological line; X_{sep}, X_{chop}, X_{dos}, X_{mix} – functions of (input) and external influence (physical- mechanical properties of mixed fodder components and conditions of external environment; Y_{sep}, Y_{chop}, Y_{dos}, Y_{mix} – resulting (output) indexes of operation of grain cleaning machine, chopper, vibrodoser and vibromixer; U_{sep}, U_{chop}, U_{dos}, U_{mix} – functions of management influence (internal regulating parameters of mixed fodder preparation machines); Z_{sep}, Z_{chop}, Z_{dos}, Z_{mix} – functions of condition above mentioned mixed fodder preparation machines (internal unregulating parameters); E_1 – quality of preparing mixed fodder in the conditions of agricultural enterprises; E_2 – energy consumption of mixed fodder preparation process

The structural scheme of technological process of mixed fodder preparation process is suggested taking into account of functional system of mixed fodder machines and peculiarities of action of each separate operation using intensified attachments of vibrating and percussive action (fig.1).

On the basis of structural scheme (fig. 1) the function of technological process of mixed fodder preparation is written as follows:

$$E = f(X, Z, U), \tag{1}$$

where E – vector-function, parameters defined productive indexes of technology of mixed fodder preparation in the conditions of agricultural enterprises (E_1, E_2); X – vector-function of external influences on grain cleaning machine X_{sep}, chopper of percussive principle of action X_{chop}, vibrodoser X_{dos}, vibromixer X_{mix} (physico-mechanical properties of mixed fodder components and conditions of external environment); Z – vector-function of technical means conditions, intensifying technological process of mixed fodder preparation (internal unregulated parameters of feed preparing machines); U – vector-function influences technical means for mixed fodder preparation in the conditions of agricultural enterprises (internal regulated parameters of feed preparing machines).

$$X= (X_{sep}, X_{chop}, X_{dos}, X_{mix})$$
$$Z = (Z_{sep}, Z_{chop}, Z_{dos}, Z_{mix})$$
$$U= (U_{sep}, U_{chop}, U_{dos}, U_{mix}).$$

After solution of the tasks on determination of parameters vector – function external influences regulated and unregulated parameters of technical means of operations fulfillment on mixed fodder preparation for function building $f(X, Z, U)$ productive indexes of operation are defined:

- grain cleaning machine: $Y_{sep} = f_{sep}(X_{sep}, Z_{sep}, U_{sep})$.
- chopper of percussive principle of action: $Y_{chop} = f_{chop}(X_{sep}, Z_{chop}, U_{chop})$.
- vibrodoser: $Y_{dos} = f_{dos}(X_{chop}, Z_{dos}, U_{dos})$.
- vibromixer: $Y_{mix} = f_{mix}(X_{dos}, Z_{mix}, U_{mix})$.

Consequently, the total function of technology for mixed fodder preparation in the conditions of agricultural enterprises using intensified attachments of vibrating and percussive principle of action is written as follows

$$E = f(f_{mix} (f_{dos} (f_{chop} (f_{sep} (X_{sep}, Z_{sep}, U_{sep}) Z_{chop}, U_{chop},) Z_{dos}, U_{dos}) Z_{mix}, U_{mix})). \tag{2}$$

Change of resulting indexes both separate feed preparing machines and whole feed preparing process will take place in the result of changing of external influences regulated and unregulated parameters.

In grain cleaning machine for separation (cleaning) of grain material (grain fodder) input parameters G (specific grain load kg/sm^2) and E_1 (initial quality of grain fodder with impurities (Z), % and moisture (W),%) influence on the plane screen with oblong openings set at an angle, output indexes (variable) are q quantity of treated grain fodder on screen of grain cleaning machine depending upon completeness of separation (ε), productivity (Q), estimated through the ratio of usage screen length (n) and probability of grain passage through the holes (P), its quality E_1^{sep} and specific energy consumption N_{sep} (or E_2^{sep}).

Under the influence on the chopper of percussive principle of action of two variables q_{sep} and E_1^{sep} output parameters are specific energy consumption N_{chop} (or E_2^{chop}), quantity (Q_{chop}) chopped particles of grain material and its quality (E_1^{chop}) as output effect, defined the alignment granulometric content (λ, M), depending upon angles jamming χ_1 and χ_2 of chopping stages. Value E_1^{chop} and E_2^{chop} are input parameters for vibrodoser.

Along with quantity (Q_{dos}) of dosing grain material according to recipe of mixed fodder preparation and specific energy consumption N_{dos} (or E_2^{dos}) for vibrodoser which includes unevenness of ingredients dosing estimating by the ratio variation C_v, meeting the zootechnical requirements and reflecting quality of each dosing
mixed fodder ingredients E_1^{dos}.

All above mentioned output indexes variables of vibrodoser (E_1^{dos} and E_2^{dos}) are parameters input parameters –amplitude (A) and frequency oscillations (ω) of vibromixer attachment, tilt angle of trough (α) to the horizon and quantity of mixing elements (n) of conical form.

Output parameters of vibromixer and whole technological line of mixed fodder preparation in the conditions of agricultural enterprises are quantity (productivity (Q), quality of mixed fodder($E_1 = E_1^{mix}$) expressed through uniformity of mixed fodder (θ), power consumption of mixing process E_2^{mix} and the whole technological line of mixed fodder preparation E_2.

It should be noted that power consumption of mixed fodder preparation process in the conditions of agricultural enterprises makes:

$$E_2 = E_2^{sep} + E_2^{chop} + E_2^{dos} + E_2^{mix} \tag{3}$$

In the result of analyses of scheme model of mixed fodder preparation follows that controlled output parameters under the influence of two input sizes G and E_1 is E_1 (or θ) and E_2. These parameters characterize output indexes of whole technological line of mixed fodder preparation in the conditions of agricultural enterprises.

In order to increase the quality and to reduce power consumption of mixed fodder preparation in the conditions of agricultural enterprises we used method of casual search according to transition from previous stage E_{n-1} to next the step is taken $j \cdot \vec{\xi}$, where $\vec{\xi}$ – single vector, indicated the direction of changing parameters intensified attachments of feed preparing machines; j – size of the step.

Proceeded from the requirements of increasing the quality and reduction of power consumption of technology of mixed fodder preparation in the conditions of agricultural enterprises using intensified attachments of vibrating and percussive principle of action, mixed fodder technological process will be carried out according to the interactive scheme;

$$E_n = E_{n-1} \begin{cases} j \cdot \xi_n, \text{ if } J \cdot \left(E_{n-1} - j \cdot \xi_n\right) < J \cdot \left(E_{n-1}\right) \\ 0, \text{ if } J \cdot \left(E_{n-1} - j \cdot \xi_n\right) \geq J \cdot \left(E_{n-1}\right) \end{cases} \tag{4}$$

where J – functional of increasing the quality of mixed fodder preparation in the conditions of enterprises J (E) → max (increasing the productivity and separation of grain heap; increasing uniformity of granulometric content; increasing the effectiveness of vibromixing and vibrodosing of mixed fodder components).

Last indexes changing of mixed fodder preparation technology according to the suggested interactive scheme (4) will be carried out by optimization parameters suggested feed preparing machines with intensified action.

The main criterions of technical efficiency of process are quality indexes of prepared product and specific energy cost.

The analyses of researches and practice show that mixed fodder production in the conditions of agricultural enterprises on higher level is not possible due to low quality of typical process of raw material treatment (cleaning, chopping, dosing and mixing), defined technological efficiency of whole mixed fodder production.

For realization of suggested technology of mixed fodder preparation in the conditions of agricultural enterprises during the fulfillment of each operation it is rationally to use intensified attachments of vibrating and percussive principle, providing maximum quantitative indexes of process with minimum energy consumption.

Thus, prototypes of new mixed fodder preparation machines of vibrating and percussive principle, included in technological scheme of mixed fodder aggregate are created during the investigations carried out in agroengineering department of the Omsky SAU [12]. Mixed fodder aggregate includes experienced prototypes of mixed fodder preparation machines of intensified principle of action for vibrodosimetering, chopping and vibromixing of mixed fodder ingredients passed the production test and applied in the form of independent machines in different agricultural enterprises including peasant and farm holdings.

We suggest technological scheme of small-size mixed fodder aggregate for production of bulk feed mixture directly in the conditions of agricultural enterprises, from feed grain crops of own production: wheat, barley, oats, peas and purchased protein- vitamin supplements (fig. 2).

Technological process of loose mixed fodder preparation includes the following operations:

- reception and treatment of starting raw material;
- feeding and distribution of components in bin sections of multicomponent vibrodosimeter;
- volume continuous and simultaneous dosing of all mixed fodder ingredients;
- chopping of ingredients in general purpose chopper with simultaneous premixing;
- final mixing of all without exceptions components in vibromixer;
- feeding of prepared mixed fodder to the bin of final production;
- storage and extradition of mixed fodder;
- Small- size mixed fodder aggregate includes multicomponent vibrodosimeter 1, bin divided into sections for different mixed fodder ingredients, two extreme sections for microsupplements, not requiring chopping.

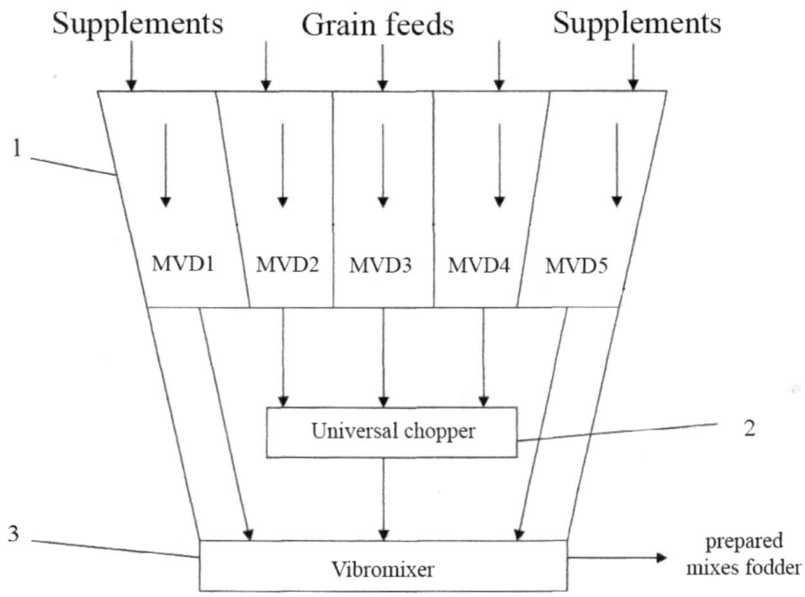

Figure 2 – Suggested technical schemes of small-size mixed fodder aggregate with intensified attachments of vibrating and impacting action

Universal chopper 2, is set under the multicomponent vibrodosimeter which allows to chop feed grain of different moisture. Prepared grinding product has lined granulometrical content at lower energy cost and the least dust-like fraction (less 5%). Mixed fodder aggregate is supplied with vibromixer 3, which is set under the universal chopper. Mixing elements with different form of working surface are set one by one inside vibromixer section. They allow to reflect bulk mixture particles with different rates and directions in the result of connection with mixing elements that promote to get feed mixture of high uniformity.

Small- size mixed fodder aggregate operates as preliminarily cleaned starting grain components from impurities from the bin section of the multicomponent vibrodosimeter 1 are delivered in predetermined ratio to the universal chopper, where subjected to grinding. Microsupplements and chopped components are delivered to the vibromixer 3, where under the influence of mixing working attachments and mixed in ready mixed fodder of high uniformity (95-97%) at the least energy consumption. Prepared mixed fodder is fed to the bin for shipping to consumers.

IOP Conf. Series: Materials Science and Engineering **221** (2017) 012020 doi:10.1088/1757-899X/221/1/012020

Thus, suggested innovative equipment and technology for the production of mixed fodder in the form of small- size mixed fodder aggregate provides effective process of bulk feed mixture preparation at the least energy consumption. It is achieved by using intensified attachments of vibrating and percussive principle of action, according to energy consumption technology. The novelty of technical solutions of mixed fodder aggregates is protected by patents of the RF on inventions and useful models. Due to its reliability, economy, simplicity in maintenance, mixed fodder aggregate can be used in small (IP, KH, KFK and others) as well as big agricultural enterprises. Model line-up suggested mixed fodder aggregate with different productivity (from 0.1-0.2 to 2-3 t/h) for commodity producers is provided for.

Conclusions

1. It is proved by scientific researches and practice that increasing the productivity of cattle and poultry depends upon the quality of feeds, and that is why it is necessary to process feed grain more rationally and efficiently in mixed fodder in the places of direct consumption.
2. Structural and technical schemes of small mixed fodder aggregate with intensified attachments of vibrating and percussive principle of action for preparation of bulk feed mixture in the conditions of big and small enterprises of agro-industrial complex were developed.
3. The usage of suggested mixed fodder aggregate allows to get mixed fodder of high-grade uniformity (95-97%) at the lowest energy costs and to reduce costing of production that becomes economically rational maintenance of livestock breeding.

References

[1] Sabiev U.K. Intensification of technological processes of preparation mixed fodder in the conditions of agricultural enterprises. Abstract of dissertation on scientific degree of Doctor of Technical Sciences, Barnaul 2012. p. 43.
[2] Yatsunov A.N. Bases of the main parametres of the vibration mixer of bulk feed with mixing elements of conical form. Abstract of dissertations on scientific degree of candidate of Technical Sciences, Novosibirsk 2008. p. 19.
[3] Sabiev U.K. Small size mixed feed aggregate // Perfection of machines and equipment in agriculture of the Western Siberia: Collection of research papers / OSAU, Omsk, 2001. p. 45-48.
[4] Kropp L.I. and others Usage of equipment UMK-F-2 (AKM-1) for production of mixed fodder on farms (recommendations). – Moscow, Agropromizdat, 1989. 25 p.
[5] Aggregate mixed-feed KA-4.Prospekt. – Authors Izaak B.I., Shmakova S.P. and others. Novosibirsk, 1988. p. 4
[6] Fedorenko I.Ya., Pirozhkov D.N. Vibrating grain layer in agricultural technology: monography / I.Ya. Fedorenko, D.N Pirozhkov. Barnaul: Izd-vo ASAU, 2006. p.166
[7] Fedorenko I. Ya. Technological processes and equipment for preparation of feeds: manual / I. Ya. Fedorenko. M.: FORUM, 2007. p. 176
[8] Sergeev N.S., Nikolayev V.N., Litash A.V., Zyazev E.V., Gajnullin E.N. Technology and technical means for preparation of bulk feed mixture on the base of multicomponent vibrating doser / Izvestiya of Orenburg SAU. 6(50). 2014. p. 68-72.
[9] Patent for invention 255526 Russia. MKI 'A 23 № 17/00, small-sized mixed fodder aggregate / Sabiev U.K. Publ. in the B.I. 25, 2000.
[10] Frolov K.V., Vibration - friend or enemy? Nauka, 1984. p. 144
[11] Dimentberg F.M., Frolov K.V., Vibration in technique and human. Znanie, 1987. 160 p.
[12] Sabiev U.K, Yatsunov A.N., Sabiev I.U. Mixed fodder aggregate / scientific research and their and practical application. The current state and ways of development. 2011. Collection of research papers Sworld. T.2. Technical sciences. Odessa: Chernomorje, 2011. p. 63-65.

IOP Conf. Series: Materials Science and Engineering **221** (2017) 012021 doi:10.1088/1757-899X/221/1/012021

Development and Analysis of Models for Handling the Refrigerated Containerized Cargoes

A Nyrkov, L Pavlova, V Nikiforov, S Sokolov [a], V Budnik
Admiral Makarov State University of Maritime and Inland Shipping, 5/7, Dvinskaya str., St. Petersburg,198035, Russia

E-mail: [a] sokolovss@gumrf.ru

Abstract. This paper considers the open multi-channel queuing system, which receives irregular homogeneous or heterogeneous applications with an unlimited flow of standby time. The system is regarded as an example of a container terminal, having conditionally functional sections with a certain duty cycle, which receives an irregular, non-uniform flow of vessels with the resultant intensity.

1. Introduction

Development of the Russian sea transport is determined not only by geography, but also by its economic advantages. Marine transportation remains the cheapest way of transport after the pipeline that leads to the demand and popularity of this mode of transport among others. Since the cost of water transport is very low, it is competitive, despite the low delivery rate.

Now the sea port is a large transport hub including a variety of transport modes. The port operation is one of the most important trends in development of the public sector of economy. It covers more than 50% of the total cargo turnover. The positive dynamics in increase of the cargo turnover of the Russian sea ports proves this fact. According to statistical data the cargo turnover in Russia increased for 4.4% for January-May 2015 as compared with the same period of 2014, and for 2014 it increased for 5.7% as compared with the previous year [1-7].

Container terminals become more common where large-tonnage refrigerating food containers are handled. Refrigerating containers as an achievement of modern technology is intended for transport of goods requiring the maintenance of special temperature and humidity conditions. Basically, these containers are used for shipping and storing perishable goods. Refrigerated containers can be used to transport virtually by all modes of transport and to store perishable goods both on the temporary basis, while transshipping in cargo terminals and for a long time. The use of refrigerated containers can deliver cargo "door-to-door" keeping the same temperature in transit and during loading and unloading when changing the modes of transport.

2. The status of the problem

While designing the complex systems for servicing the container terminals it is reasonable to apply complexes of coordinated and information-consistent and polynomial models, which are determined on the basis of the active identification methods. The current theory of complex systems identification is inextricably linked with the theory of optimal control theory and optimal experiment theory. Nowadays, there is a large number of papers devoted to optimal identification systems in various classes of models. At the same time, if at the initial stages of development of the identification theory, methods of optimal control theory were used in general, now it is a trend towards a wider use of the optimal experimental design theory methods.

Development of a complex of polynomial and classical models of systems for servicing container terminals requires special plans for the computational experiment, which take into account specific features of the systems and processes of their designing. Planning of the computational experiment based on specific computational models and processing of the results, in accordance with the criterion of optimality, allows for the active identification of handling and storing the refrigerated containerized cargoes, i.e. getting a set of consistent models representing polynomial relationships of various indices of the process quality with the parameters of the systems under review.

To cut costs at the stage of the cargo transportation to the port and to reduce the provisional time of delivery, container transportation is widely used, as they have a very high degree of security. Modern containers are technically designed and manufactured in compliance with all the rules and standards in force, which ensure their strength and integrity. This is one of the most convenient and popular forms of transportation. In the world the popularity of using containers as a universal receptacle for transportation of goods is growing.

The technical equipment in the ports in respect of import and export, cannot always be ready for storage or handling the containerized cargo. In this regard, it is required to develop models optimizing and reducing the cost of equipment and resources [5-11]. Let us consider the mathematical model, by means of the queuing system, with which it is possible to reduce the error in the calculation of probability characteristics.

3. Scientific novelty

Reasoning from theoretical premises, the main consequence of nanopowder introduction into the melt should be refinement of the macro- and microstructure, as the powder particles must serve as nuclei of new grains.

When using the Markov's model queueing system, the exponential service law is assumed. When calculating the QS probability characteristics, the use of the exponential law can lead to errors and lack of precision.

Deterministic approaches to describe the capacity of container terminals, in fact, do not reflect the specific features in processing and storing refrigerated goods, and in particular, do not take into account the time interval between the arrival of requirements in the service channel and release requirements of the channel, however, the delivery of batches of containerized cargo to the terminal is a random flow of events. Under real conditions the operation of the terminal infrastructure in terms of cargo handling processes is not adequate to the above mentioned assumptions [2]. Given the specificity of container terminals to meet the challenges of research and optimization, it is required to consider random processes in handling the vessels, describing these processes based on probability models [3]. Let us consider a different approach to the calculation of the characteristics of the QS probability model.

A mathematical model of the storage process allows for the irregular arrival of consignments and the random time of their finding in the stock within refrigerated container terminals. That leads in some cases to the demurrage and excess storage spaces in warehouses, in other cases it leads to the formation of queues.

In order to solve the problem of research and to develop a probability model it is necessary to investigate random processes in cargo storage and make their probability analysis. To move from one state to another, data of the cargo storage processes in certain moments of time are used. Transition of the process takes place in moments of change in cargo handling state, i.e. when a new vessel is entering the terminal, one of the berths is becoming free. The number of vessels in the queue for storing goods changes as well.

To reflect the major events of the system let us plot the process of functioning of an open multi-channel queuing system graphically. Fig.1. shows Multichannel queuing structure. Fig. 2 shows the basic processes of the multi-channel open QS.

IOP Conf. Series: Materials Science and Engineering **221** (2017) 012021 doi:10.1088/1757-899X/221/1/012021

Describe every event in the system under modelling:

1. Arrival of vessels in the terminal.
2. Entering the vessels to the roadsteads (queue / loading area).
3. Determination of the conventionally-functional section for servicing the vessel.
4. Waiting for emptying one of the conventionally-functional sections.
5. Leaving the loading area by the vessel.
6. Time of servicing the vessels in the conventionally-functional section.
7. Emptying the conventionally-functional section.

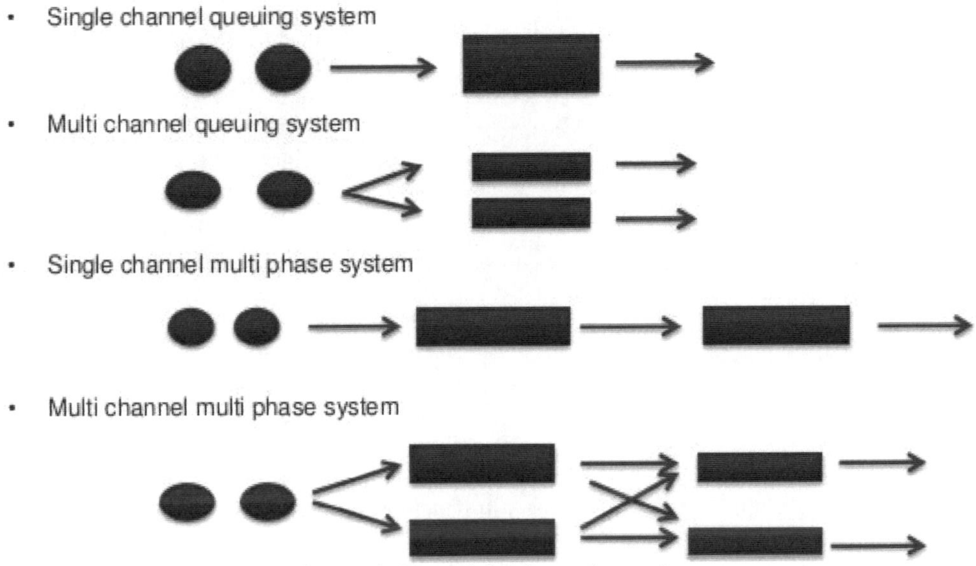

Figure 1. The Multichannel queuing structure.

Leaving the terminal by the vessel.

Let us introduce the notation of the refrigerated container terminal with F conventionally functional sections with the same duty cycle φ, which receives the irregular flow of the vessels m (a number of vessels) with the resultant intensity λ. The resultant intensity of the vessels' arrival in the terminal is determined by the intensity and a number of vessels in the queue.

If the vessels are similar in terms of their service, it is believed that a uniform flow of applications or claims (vessels) enters the QS, i.e. all applications for discharging are the same [4]. In previous papers queuing systems based on probability models were considered in relation to the uniform flow of applications, namely all incoming vessels have the same specification, i.e. the same capacity and size of container consignments (number of containers on board the vessel).

However, there may be certain non-uniformity of the vessels' flow, namely vessels may have different characteristics: the number of containers on the vessel, and capacity of containers, etc.

Assuming that the flow of consignments (vessels) coming into the port obeys the Poisson distribution law, it may be considered the simplest one (stationary Poisson law), which has three following properties: ordinariness, lack of aftereffect and stationarity.

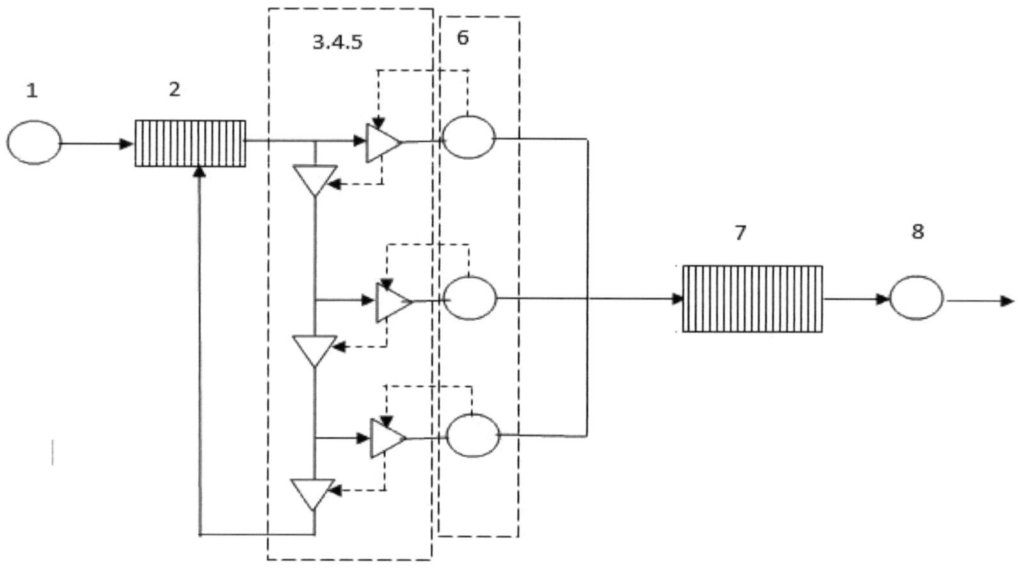

Figure 2. The process of functioning of the multi-channel open queueing system.

4. Results and development of algorithm

Using polynomial and mathematical models schemes of the functioning processes and conditions of multi-channel open QS described above, the probability characteristics were calculated by the software.

The choice to use this tool as a software development, allows us to solve all of the above tasks, and provides a simple and convenient programming, and the possibility of future upgrades. The developed software is a set of libraries of functions and the dynamic content aimed at calculating the probability characteristics of the storage and handling reefer container terminals. The program provides the capability to input a parameter with no quantitative restrictions such as the type of vessel. The size of consignments, the size (capacity) of the container, the fill rate and other characteristics for a particular type of vessels are taken into account for every type of vessel.

In the course of development the following tasks have been set and solved:

- implementation of initial data input: F is the number of conventional functional sections; ms are vessels of s-type, where s is the number of "non-uniform" vessels;
- φs - the duty cycle (Figure 3).
- entries must comply with the conditions and they should be checked for correctness;
- calculation of the average number of consignments in the queue for storage;
- the intensity of the arrival of batches of containers;
- the average time waiting for the container batches to be stored.

According to the results of calculations it may be stated that the average waiting time by increasing the duty cycle significantly increases and decreases by increasing the number of quasi-functional sections in the refrigerated terminal. Calculations have shown that by using *rs* and *rcp* indices differ slightly. It proves that it is possible to use average value *rcp* in further calculations.

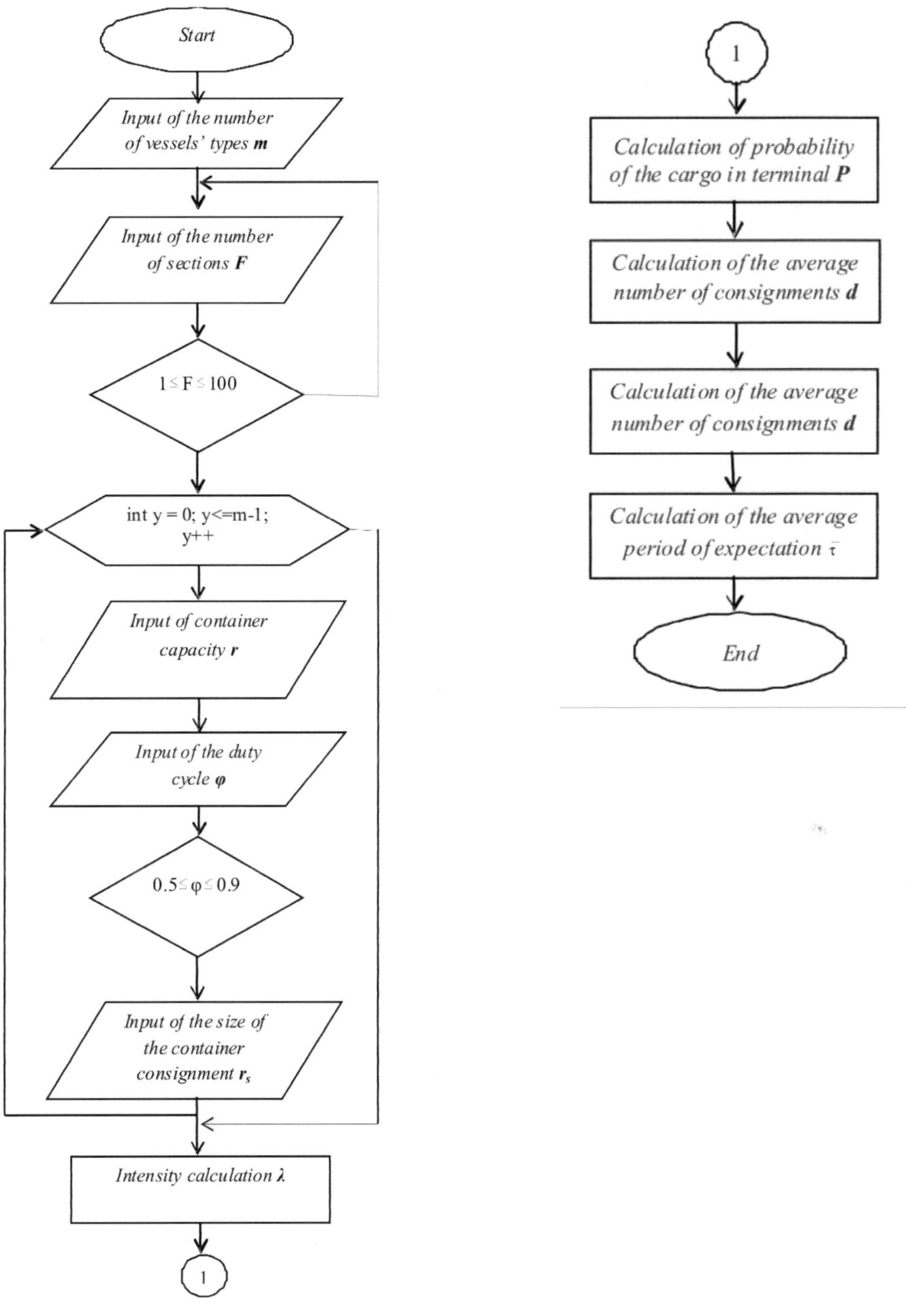

Figure 3. Block diagram of the calculation of the probability characteristics of the storage and handling of reefer containerized cargoes.

With the development of the terminal a new organizational structure was implemented, at the same time methods and tools for assessing the transshipment terminal and other parts of the multimodal transport were improved [5].

In order to simulate a model of loading-unloading terminal operation the queueing system tools are used. Open QS queuing systems are the basis of classical and polynomial modeling of cargo handling [6]. To determine the characteristics of the processes in handling the container cargoes, it is not always appropriate to apply the available queuing model, as these models do not correctly describe the processes of the port operation [7-8].

The deterministic methods to describe the capacity of container terminals, in fact, do not reflect the specific features in handling and storing refrigerated cargoes, in particular, such as the time of delivery of the container consignments to the terminal is considered as a random flow of events [15-20].

5. Conclusion

Thus, the resulting probability model allows us to analyze the storage processes in the refrigerated container cargo terminals, taking into account the specific features of cargo handling. While calculating the mathematical model, we can find the following parameters: the average number of container consignments in a queue for storage; average reduced expectation time, the intensity of the arrival of every batch of containers in the terminal and the fill factor of the conventionally functional sections. The results are more accurate than using the "standard" deterministic method, which does not fully reflects the specific features of handling the reefer container cargo due to the fact that using this method does not take into account such important aspects as the irregular flow of events, a random time of stay of vessels in the terminal (in the conventionally functional section) and a random variable handling time.

References

[1] http://www.morport.com/rus/

[2] http://www.wiraj-perevozki.ru/text_7.html

[3] Rusinov I A The formalization and optimization of processes of handling the refrigerated cargoes at specialized terminals Polytekhnika 472

[4] Galin A A 2016 The impact of restrictions on the generic simulation model of port development, Newsletter of the Admiral Makarov State University of Maritime and Inland Shipping 1 7-14

[5] Lovyanikov D S, Zubarev Ya 2014 *Newsletter of the Admiral Makarov State University of Maritime and Inland* **1** 109-113

[6] Zubarev Yu, Gainullin A S 2011 *Newsletter of the Admiral Makarov State University of Maritime and Inland Shipping* **1**

[7] Chernyi S 2016 *IOP Conference Series: Materials Science and Engineering* **124** 012048.

[8] Chernyi S 2016 *Transport and Telecommunication Journal* **17(3)** 202–211 DOI: 10.1515/ttj-2016-0018

[9] Semenov K M 2014 Planning of cargo handling in sea ports and terminals based on discrete-eventual simulation Kaliningrad 171

[10] Cinquini C, Venini P, Nascimbene R, Tiano A 2001 *Structural and multidisciplinary optimization* 3-22

[11] Du X, Zhang P, Ma W 2016 *Journal of Computational and Applied Mathematics* **305** 92-114

[12] VieiraG, Neto F, Ribeiro J 2015 *International Journal of e-Navigation and Maritime Economy* **2** 73-86

[13] Keshavarzzadeh V, Meidani H, Tortorelli D 2016 *Computer Methods in Applied Mechanics and Engineering* **306** 47-76

[14] Nyrkov A P, Belousov A S, Sokolov S S 2015 Modern Applied Science **9(5)** 162–176

[15] Podevins O 2007 *International Symposium on Logistics and Industrial Informatics* 151-155

[16] Löhndorf N 2016 European Journal of Operational Research **255(1)** 121-132

[17] Budnik V, Chernyi S 2016 *Procedia Engineering* **150** 2150-2156

http://dx.doi.org/10.1016/j.proeng.2016.07.256
[18] Mclaughlin S, Pereyra M, Hero A, Tourneret J, Pesquet J 2016 IEEE Journal of Selected Topics in Signal Processing **10(2)** 221-223
[19] Renders J M, Flasse S P 1996 *IEEE Transactions on Systems, Man and Cybernetics, Part B (Cybernetics)* **26(2)** 243-258
[20] Zhilenkov A, Chernyi S 2015 Procedia Engineering **100** 1247-1252 http://dx.doi.org/10.1016/j.proeng.2015.01.490

IOP Conf. Series: Materials Science and Engineering **221** (2017) 012022 doi:10.1088/1757-899X/221/1/012022

Systems of Geo Positioning of the Mobile Robot

M V Momot [a], A V Proskokov, D N Nesteruk, M Ganiyev, A S Biktimirov

Yurga Institute of Technology, TPU,
652055, Russia, Kemerovo Region, Yurga, Leningradskaya St., 26

E-mail: [a] mikl_1239@mail.ru

Abstract. Article is devoted to the analysis of opportunities of electronic instruments, such as a gyroscope, the accelerometer, the magnetometer together, the video system of image identification and system of infrared indicators during creation of system of exact positioning of the mobile robot. Results of testing and the operating algorithms are given. Possibilities of sharing of these devices and their association in a single system are analyzed. Conclusions on development of opportunities and elimination of shortcomings of the received end-to-end system of positioning of the robot are drawn.

Introduction

In modern economy special attention is paid to problems of transfer to robots of routine work, and also the works connected with the increased danger. It is for example carrying out field works, navigation in conditions of increased content of toxic agents. If to consider only agricultural machinery, then the increasing application is found by automatic control systems the movement and systems of parallel driving. Similar systems include the electronic intellect consisting of a mathematical apparatus and system of sensors and the executive mechanism allowing to transform solutions of electronic intellect to the relevant movements of the robot. The robot can independently move, but be under control of the operator. The operator can interfere with operations of the robot in situation which cannot independently process by the robot. For example, the robot does not learn the area, or the robot noticed unfamiliar objects. Use of robots during the field works allows to accelerate significantly processing of the soil and to lower costs of the energy carrier, including due to decrease in mass of the robot.

The majority of systems of automatic piloting use possibilities of satellite positioning of GLONASS or GPS [1, 2, 5]. With different degree of success, attempts of use of systems of machine vision [3, 4], and also the end-to-end systems of positioning including the different principles [6] are made. However It should be noted that in spite of the fact that high-precision systems of satellite navigation allow to position the robot with an accuracy of several centimeters use them is not always justified from the financial and technical points of view. Moreover use of systems of local positioning is a good opportunity for development of the mathematical and technical apparatus.

Proposed solutions

At development of the system of positioning it is necessary to consider hardware features of each of the making subsystems. The skeleton diagram of the received system of positioning is submitted in the figure 1. Actually the following blocks are engaged in data collection: *F_images()* image identification subsystem, *F_satellites()* positioning subsystem on the basis of satellite navigation, *F_beacons()* subsystem of definition of a position on the basis of external infrared beacons, *F_inertia()*

positioning subsystem on the basis of inertial mechanisms and magnetic field of the earth, *F_communication()* subsystem of external control and coordination of the movement.

The *F_intel()* function obtains data from the subsystems *F_intel(F_satellites(t), F_beacons(t), F_images(t), F_inertialt(t), F_communication(t))* and makes a decision on management of the executive bodies of the robot, proceeding from an objective and indications of subsystems of information collection.

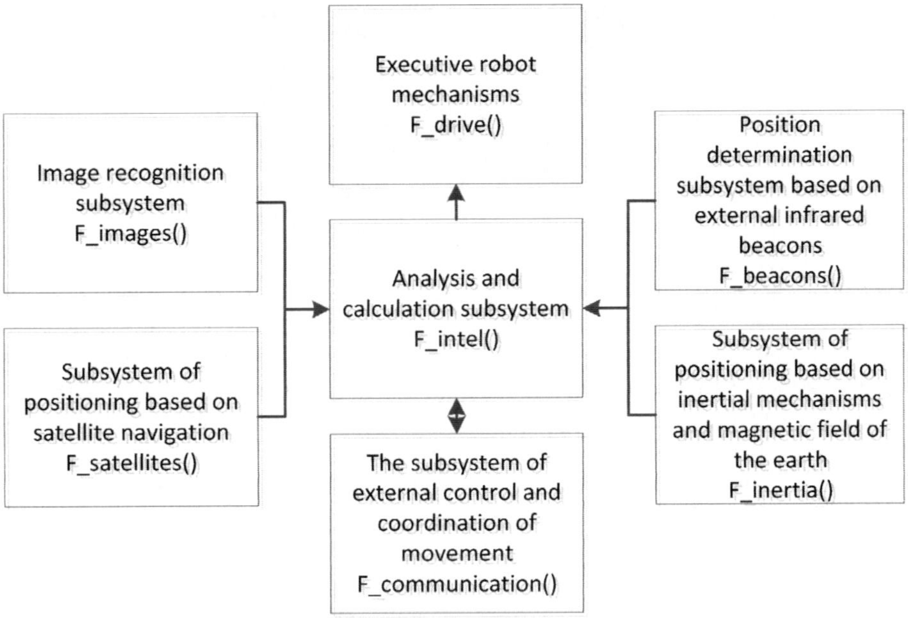

Figure 1. Structure diagram of an end-to-end system of positioning of the robot

The image identification subsystem involved on the robot is based on use of library of computer sight of OpenCV. The program processing information through OpenCV functions on the single board computer RaspberyPI which is equipped with system of video cameras. The method of recognition of graphic images on the method of Harris Corner Detection [3, 5]. The program is able to define standard obstacles, such as trees, bushes, columns, large animals and people, to select unfamiliar objects, for example, the equipment thrown in the field. Anyway information is transferred to the Subsystem of the analysis and calculations which should make a decision on processing of a regular or emergency situation. Information is transferred in the look represented in the figure 3, surely includes distance to an object, the object sizes, an object arrangement concerning the robot, the speed of its movement and a motion vector. Of course, these data cannot be considered completely reliable, they are in addition analyzed and compared to information from other subsystems and the internal map put in the robot. In necessary cases additions are made to the internal map. Use of several spaced cameras gives the chance to measure distance to objects and there's size. The history of coordinates uses to measure the speed of objects, relative in relation to the robot.

Figure 2. Matching interest points of graphic images on the method of Harris Corner Detection

It is worth noticing that the pattern recognition system has no idea of robot motion speed, but can take part in its measurement on a measurement base of characteristics of motionless objects. In other words, if it is known that an object is not mobile, but its characteristics (the size and motion vector) changes, there forms a basis for calculation of the robot speed.

Information is transferred in a matrixes I_j, where j - number of the found object.

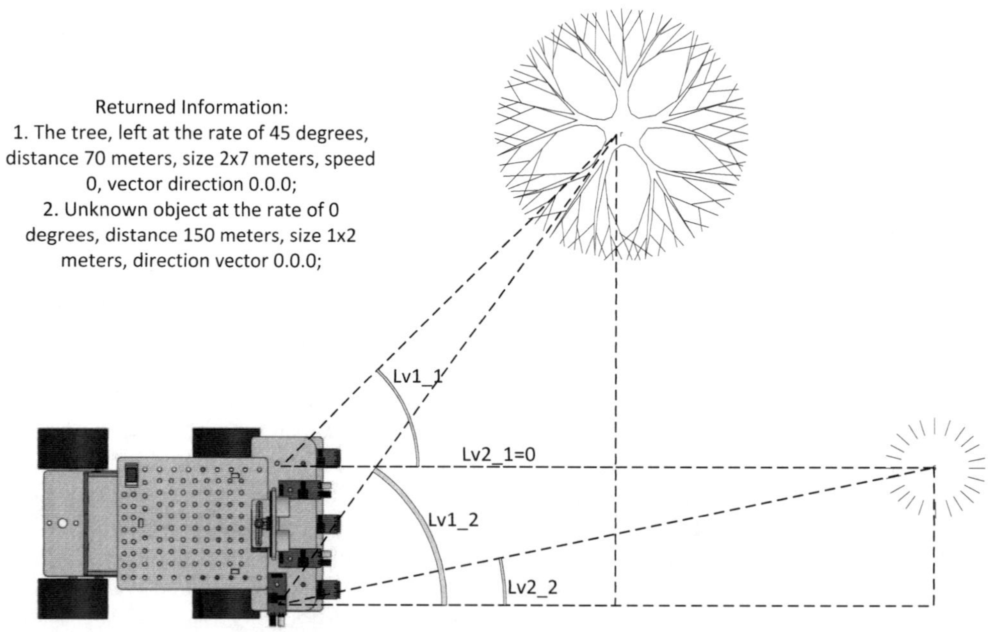

Figure 3. Work of a subsystem of image identification

The subsystem of a position definition on the basis of external Bluetooth and RFID was in details described by us in [4]. On perimeter of a field the infrared frequency-modulated beacons differing in the coded signal are installed. On the robot the system of rotary infrared sensors is installed. Corners between a motion vector of the robot and the direction to indicators are determined by sensors, and by

IOP Conf. Series: Materials Science and Engineering **221** (2017) 012022 doi:10.1088/1757-899X/221/1/012022

the use of several sensors distance is calculated (fig. 4). Such system gives the chance to lay a route across the field and rather precisely to adhere to it.

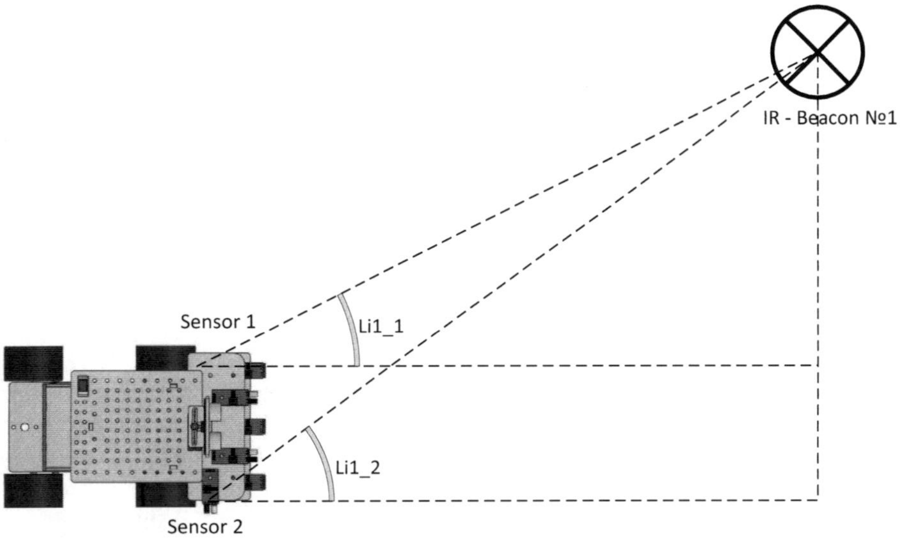

Figure 4. Robot and RFID (determination of coordinates of the beacon)

The values calculated in this subsystem include the following fields:

1. The number of the found beacons;
2. Identifiers of beacons which include the individual numbers of coded by a frequency modulated signal;
3. The distance to each of the found beacons calculated by the principle of stereosight due to use of two or more sensors of an infrared modulated signal;
4. Corners to all found beacons.

Information is transferred in a matrix IB_k where k - number of the found object.

The positioning subsystem on the basis of inertial mechanisms and magnetic field of the earth includes at least on one electronic instrument of the following type: gyroscope, accelerometer, compass/magnetometer. The data obtained from devices are processed therefore the matrix Pk in which the following information forms are presented:

1. Turning angle of the robot of rather initial "null value";
2. Vector of acceleration of the robot;
3. Time of measurements;
4. Estimated error.

The positioning subsystem on the basis of satellite navigation consists of the GPS/GLONASS module and the microprocessor module of primary analysis of the acquired information. The subsystem transfers the following information:

1. Geographical coordinates of the robot;
2. Direction movement of the robot;
3. Robot speed.

Information is transferred in a matrix SN.

IOP Conf. Series: Materials Science and Engineering **221** (2017) 012022 doi:10.1088/1757-899X/221/1/012022

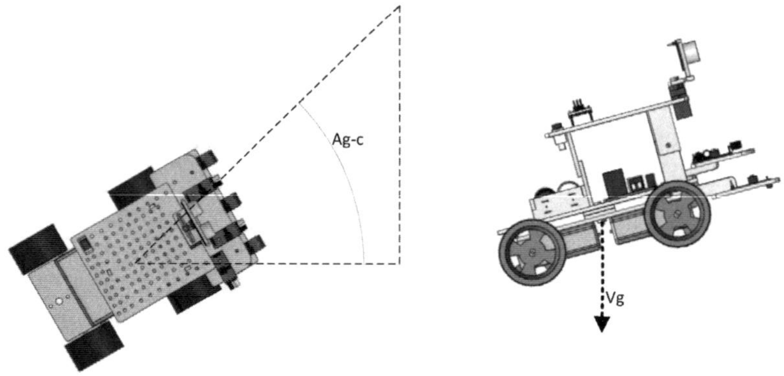

Figure 5. The robot (inertial, gravitational, magnetic sensors), where Ag-c - a turning angle of the robot, Vg - an acceleration vector

The subsystem of external control and coordination of the movement *F_communication()* can execute several functions, one of which is visual observation of the robot in the automatic mode, visual tracking of its characteristics and transfer of the calculated information to a subsystem of the analysis and calculations. In this case the block of visual observation in the form of a complex of video cameras and the microprocessor block of calculation on the basis of visual images is in a subsystem. Let's notice that this subsystem can be stationary or to have "moving bases", for example, in the form of the flying device (balloon) which can conduct simultaneous observation at once or in turn behind several mobile robots.

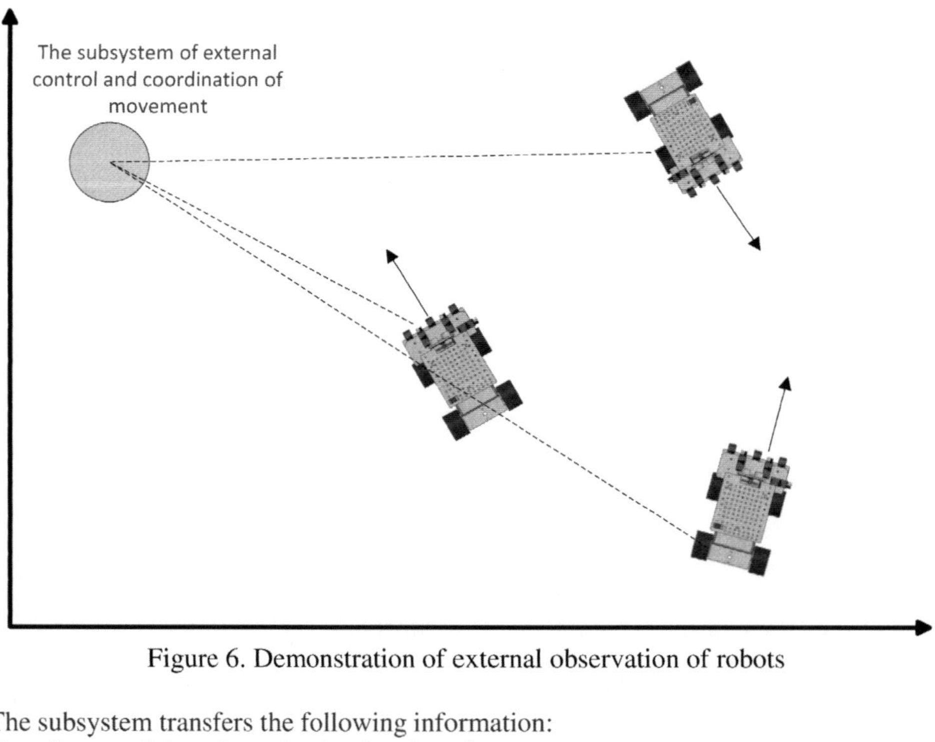

Figure 6. Demonstration of external observation of robots

The subsystem transfers the following information:

1. Coordinates of the robot of rather local section;
2. Direction movement of the robot;
3. Robot speed;
4. Probable obstacles.

Information is transferred in a matrix *EO*.

Accuracy of indications of the above described subsystems has some error, and during certain time frames of the indication can be absent or be less informative, for example, during fog, smoke, at failure of a part of the equipment – satellite system of positioning. Therefore, the algorithm for mixing data and calculating the "reliable" coordinates of the robot and the vector of its movement must be flexible enough and work with incomplete information.

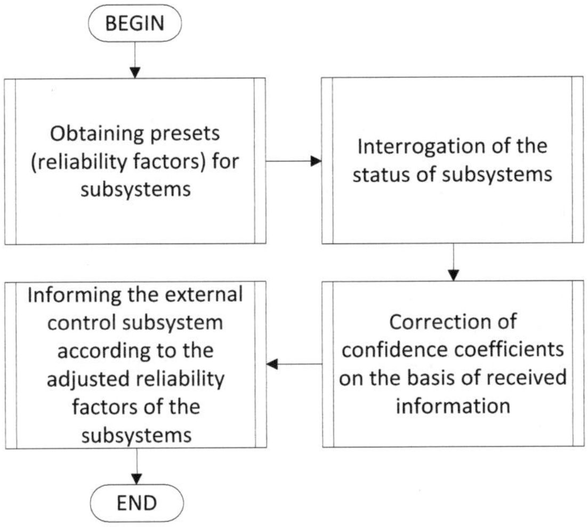

Figure 7. Formation of coefficients of subsystems reliability

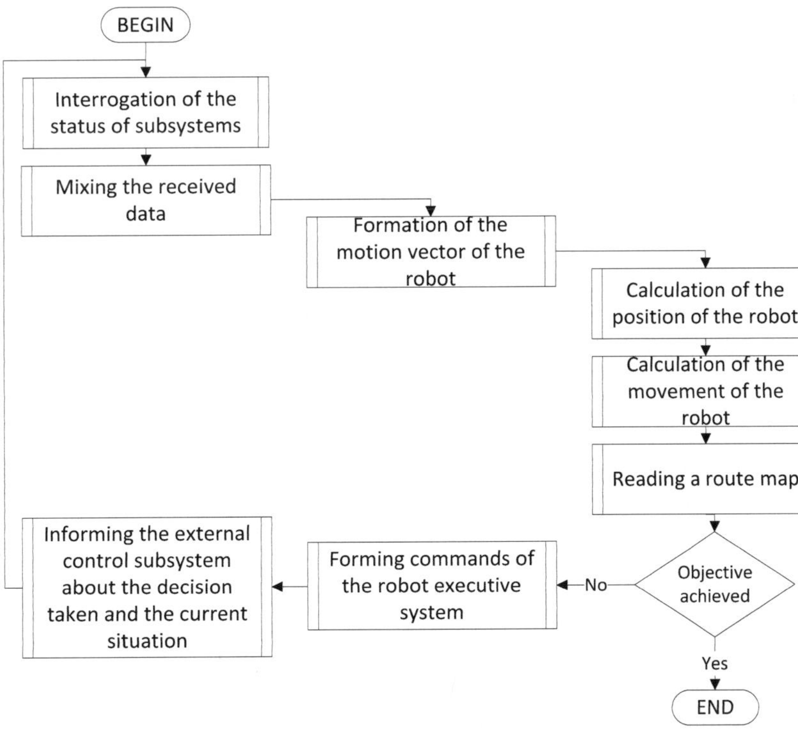

Figure 8. An algorithm of work of a subsystem of the analysis and calculations at the movement of the robot on the set trajectory

IOP Conf. Series: Materials Science and Engineering **221** (2017) 012022 doi:10.1088/1757-899X/221/1/012022

Conclusions

As a result of use of the above-stated solutions the system of positioning of robots of agricultural purpose which allows to carry out field works with high accuracy without obligatory binding to high-precision system of positioning on the basis of GPS/GLONASS is created. The system has the small cost and high rates of quality. Integration of the specified system in a complex makes possible to receive the most effective structure from the point of possession cost.

References

1. Momot, M. V., Proskokov, A. V., Natalchenko, A. S., & Biktimirov, A. S. (2016, August). Automatic Positioning System of Small Agricultural Robot. In *IOP Conference Series: Materials Science and Engineering* (Vol. 142, No. 1, p. 012104). IOP Publishing.
2. Rosten, E., & Drummond, T. (2006). Machine learning for high-speed corner detection. *Computer vision–ECCV 2006*, 430-443.
3. Royer, E., Lhuillier, M., Dhome, M., & Lavest, J. M. (2007). Monocular vision for mobile robot localization and autonomous navigation. *International Journal of Computer Vision*, *74*(3), 237-260.
4. Kulyukin, V., Gharpure, C., Nicholson, J., & Pavithran, S. (2004, October). RFID in robot-assisted indoor navigation for the visually impaired. In *Intelligent Robots and Systems, 2004.(IROS 2004). Proceedings. 2004 IEEE/RSJ International Conference on* (Vol. 2, pp. 1979-1984). IEEE.
5. Kravtsov, S. V., & Rumjantsev, K. E. (2014). Positioning autonomous mobile robot based on measurements onboard digital stereo vision system. *International Journal of Robotics Applications and Technologies (IJRAT)*, *2*(2), 37-77.
6. Emmi, L., Gonzalez-de-Soto, M., Pajares, G., & Gonzalez-de-Santos, P. (2014). New trends in robotics for agriculture: integration and assessment of a real fleet of robots. *The Scientific World Journal*, *2014*.
7. Backman, J. (2013). Navigation System for Modular Agricultural Machines using Optimal Control Methods and Industrial Standard Network.

Calculation of the Strip Foundation on Solid Elastic Base, Taking into Account the Karst Collapse

R Sharapov, N Lodigina

Vladimir State University, 23, Orlovskaya str., Murom, 602264, Russia

E-mail: info@vanta.ru

Abstract. Karst processes greatly complicate the construction and operation of buildings and structures. Due to the karstic deformations at different times there have been several major accidents, which analysis showed that in all cases the fundamental errors committed at different stages of building development: site selection, engineering survey, design, construction or operation of the facilities. Theory analysis of beams on elastic foundation is essential in building practice. Specialist engineering facilities often have to resort to multiple designing in finding efficient forms of construction of these facilities. In work the calculation of stresses in cross-sections of the strip foundation evenly distributed load in the event of karst. A comparison of extreme stress in the event of karst and without accounting for the strip foundation as a beam on an elastic foundation.

1. Introduction

For construction of residential and public buildings are often complex are removed from the point of view of the geological structure and processes of the territory. One of these dangerous and poorly studied karst processes is widely distributed on the territory of Russia. The most widespread manifestations of karst processes are characterized by Vladimir, Kaluga, Kursk, Moscow, Arkhangelsk, Novgorod region, Republic of Bashkortostan, Tatarstan and Mari El Republic [1, 2, 3].

Such rocks as gypsum, and other conglomerates, dissolve and disintegrate under running water. In them formed karsts, and on the earth's surface - flops, karst subsidence, which over time transformed into sinkholes, subsidence or subsidence [1].

Karst processes greatly complicate the construction and operation of buildings and structures. Due to the karstic deformation at different times there have been several major accidents, which analysis showed that in all cases the fundamental errors committed at different stages of building development: site selection, engineering survey, design, construction or operation of the facilities [4, 5, 6, 7].

Foundations must meet the following basic requirements: have sufficient strength, resistance to overturning and sliding in the plane of the sole to resist the influence of groundwater and aggressive treatment, as well as the impact of weather conditions (frost), comply with durability term building service, be technologically advanced in manufacturing and economical.

The design of foundations are: solid, strip, bar and pile. Primers are used as bases of buildings and constructions, divided into: clay, sand, coarse, rocky, peaty.

The most important tasks of engineering and geological research is the study of patterns of development of karst, its assessment of the current status and future growth opportunities. Based on

IOP Conf. Series: Materials Science and Engineering **221** (2017) 012023 doi:10.1088/1757-899X/221/1/012023

the research must be identified areas where karst processes do not cause and will not cause significant difficulties in the construction and constitute a danger for the construction [8, 9].

In order to solve specific engineering and construction tasks necessary to have not only the general information regarding the presence of karst and the general laws of its development within the construction site. It should also possess detailed information on microdistrict all his ties with neighboring regions, within which the projected construction.

For the calculation and design of measures to ensure the stability of the buildings necessary information about the current state of karst processes on this construction site with sufficiently accurate for engineering purposes the presence of caverns, sinkholes, underground tunnels and the like, rather than random sampling of finding the passage borehole [10].

Needed such geotechnical data on the karst, which would have created a solid basis for the calculation of measures to ensure the reliability and stability of the buildings. This is the main issue to be determined in the production of engineering and hydrogeological survey.

In the presence of data required for the design, you can confidently tackle the issue of the construction of buildings in the karst area. It is also possible resolve the question of the degree of stability and reliability in the roof strata discovered caverns, need and opportunities to strengthen the bearing capacity of the column or unloading it at the desired degree, termination of karst processes in terms of thickness of the roof voids, ensuring normal operation of buildings.

Figure 1. Karst collapse in Guatemala.

Construction in karst areas is associated with considerable difficulties, as the karst rocks are fragile base. Emptiness reduces the strength and stability of the rock foundations of buildings and structures. The development of karst forms can cause harmful rainfall or even complete destruction of structures. Karst process is especially dangerous to hydraulic structures. Karst cavities are possible leaks of water from the reservoirs, canals.

During construction in karst areas is necessary to carry out a series of measures aimed at halting the development of karst forms, increasing the stability and strength of rocks [11]:

- Protect the soluble rocks from the impact surface and groundwater;
- Reinforcing the karst rock.

For proper design of buildings and structures in karst areas requires detailed geotechnical study, which should be comprehensive. In this study the climate, vegetation, hydrology, geology, ground, groundwater and including everything related to karst forms [12, 13].

2. Calculation of the strip foundation

Design bases and foundations of buildings erected on the karst territories, carried out in accordance with the general requirements of SNIP 2.02.01-83. At the same time provide for measures that exclude the possibility of the formation of karst deformation and reduce their adverse effects on the building.

However, implementation of these measures does not always rule out the possibility of development of karst deformation, and in some cases becomes technically impossible or impractical to use them.

In this case, provide construction measures, appointed on the basis of the foundations and underground structures construction calculation, taking into account the possible development of karst deformation [14, 15].

Construction of the foundations of buildings erected on the karst areas have their own characteristics. The calculations of foundations made on the basis of forecasting the size of karst manifestations and the probability of their formation at the base of the projected building.

The calculated location of failures vary and are assigned based on the most adverse operating conditions of the foundation construction. Challenges calculation bases on the karst grounds taking into account the joint work with the above-ground structures are solved on a computer.

Known crash buildings and civil engineering development in karst areas included geotechnical hazard category. Deformation due to subsidence of the earth's surface is transmitted form of subsidence. In this superstrate structure is not cut off vertically and laterally compressed displacement into a free space. Because rock displacement towards the center of gravity of the hollow space is also formed along the horizontal movement to vertical, which may cause lengthening or shortening of the surface structures on the site location leads to the effects that the construction of a stretching or compression. These or other process elements move differently act on the structure as a whole or to its separate structure. Accordingly, the required various safety measures or special work. Uniform deposition does not create additional stresses in the construction of buildings and therefore not taken into account in the design and calculations. However, it can have an impact on the conditions of sanitation and the relative rise of groundwater to the surface, leading to hydration base facilities. In addition, there may be problems with underground utilities.

The difference in the sedimentation rate results in the slope of the base structure, which accounts for a maximum of the transition from the convex portion to the subsidence of the concave portion. In this regard, skew additional horizontal component forces arise along with the usual vertical forces which cause bending moments in building constructions. Stresses arising from bending moments depend on the flexural and bond strength structurally interconnected parts of the structure. Most buildings have some stiffness so that they can follow a curvature to some extent without damage. When more rigid support structure may be a higher concentration of stresses [16].

With the sudden formation of the failure on the ground in some parts of the structures are formed by the base spans and console. In these cases a stress redistribution and therefore uneven load distribution of structures on the substrate, wherein the load cannot be redistributed in part. In the construction having compressive and tensile stresses which can be perceived only to a certain degree, the tensile strength in excess of the formation fracture.

In static analysis of structures must be applied reasonably estimating failures diameter, while also takes into account the depth and frequency of karst. The research programs have the following objectives: a high karst-a failure risk facilities should be thoroughly protected structurally, it is necessary to conduct monitoring of possible deformations. For large lesions often more economical to dismantle the construction and start new construction [17, 18].

Theory analysis of beams on elastic foundation is of great importance in construction practice. Specialist engineering facilities often have to resort to multiple designing in finding efficient forms of construction these buildings. This primarily relates to buildings on soft ground, strip foundations reinforced concrete viaducts, floating bridges, tall foundations of knowledge, crane paths, floors of industrial buildings, concrete pavements and airfields [19]. On the construction of such facilities leaves almost half of the concrete used in construction. Such huge material and financial costs require careful attention to the issues of structural analysis on elastic foundation. The problems associated with the study of structures, lying on an elastic foundation, represent one of the most urgent, complex and interesting problems of structural mechanics. In recent years attention to these problems increasingly growing. On the one hand this is due to the pressing needs of engineering practice, and on the other the development and improvement of methods of calculation. A major role is played by modern computer technology. In its calculation model structure analysis on an elastic foundation is first set and reviewed the terms of its contact with the ground, and then calculates the [20]. Normally, structures are made of a homogeneous and isotropic material. Their contact with the ground under all combinations of external influences considered continuous, that is supposed to match the vertical movement of construction and surface of the base over the entire contact area.

In engineering practice frequent beam lying on a continuous elastic foundation. Elastic base called base beams such that the beam is deformed by the weight and the load situated on it, and thus renders the elastic sag resistance. Beams, lying on this basis, called beams on an elastic foundation. These beams railway sleepers can be attributed path strip foundations of various structures that transmit the load to the ground, the foundations of dams and others. Static equation allows only find the overall reaction on the part of the base and makes it impossible to determine the law of distribution of the reaction along the length of the beam. The value at each point of the reaction depends on the beam deflection, and the deflection of the beam in turn depends on the response from the base.

Figure 2. Collapse under the house foundation.

To solve the problem commonly accepted hypothesis linking the value of reactions to the draft base [21]. One of the most common hypothesis is the hypothesis of proportionality between the reaction and the draft of the base. Such a hypothesis regarding the properties of the soil, was first proposed N.I. Fuss in 1801 and applied to the beams on an elastic foundation used E. Winkler. According to this hypothesis base reaction at each point is proportional to the elastic precipitate at this point. In other words, the greater the deflection, the greater the resistance and reactance. Thus, the elastic base reaction with respect to the beam is a changing along the length of the beam intensity of the load:

$$q_0 = -k \cdot y, \tag{1}$$

$$k = k_0 \cdot b, \tag{2}$$

where k_0 – constant factor characterizing the rigidity of the base, called the coefficient of bed; it is reactive force per unit area of 1 cm^2, and emerging from the base at a deflection of the beam is equal to 1 cm. This ratio is determined from experiments and the dimension of its N/cm^2; b – width of the sole of the beam.

Therefore, $k = k_0 b$ represents a linear coefficient of the base. It is numerically equal to the base of the reaction per unit length of the beam at a deflection of 1 cm. The dimension k N/cm^2 coefficient.

Theory analysis of beams on elastic foundation is very detail was developed by Krylov using the method of initial parameters [20]. The advantage of this method lies in the fact that for any kind of load and any method of fastening the ends of the beam equation of the bent axis of the beam on an elastic foundation comprises four primary parameter, which is the deflection, tilt, bending moment and shear force in any cross-section of the beam, is taken as the origin. Two of these parameters are always known. To find the other two have to solve a system of two linear equations with two unknowns.

Strip foundation in this paper is considered as a beam on an elastic foundation under the action of uniformly distributed load it (fig.3).

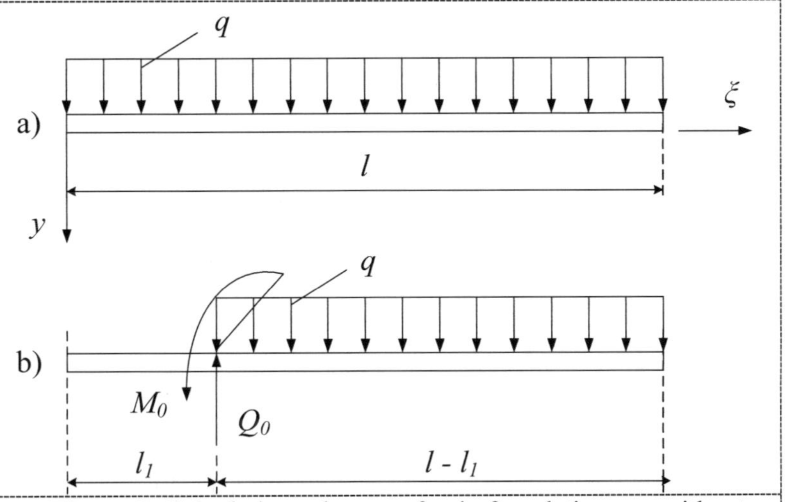

Figure 3. Calculation schemes of strip foundation: a – without considering the karst collapse; *b* – with considering the karst collapse.

In figure 3: q – the load of the weight of the structure and strip foundation; l – the length of the strip foundation; l_1 – the length of the strip foundation over karst collapse; M_0, Q_0 – initial parameters: the bending moment and transverse force of the construction and foundation weight activities over karst collapse.

Two cases are considered:

- The load acting on the entire length of the strip foundation l;
- Load acts over a length $l - l_1$ (l_1 - diameter karst failure [22]).

Accordingly, the expression for the deflection angle and the bending moment can be written as:

- Without karst failure (initial parameters $M_0 = 0,\ Q_0 = 0$)

$$y(\xi) = 91.18\cdot10^2\cdot q\cdot Y_1\cdot(l - 10.95\cdot10^5\cdot\xi); \quad (3)$$
$$\varphi(\xi) = 33.3\cdot q\cdot Y_4\cdot(10.95\cdot10^2\cdot\xi - l); \quad (4)$$
$$M(\xi) = 10.95\cdot10^2\cdot q\cdot Y_3\cdot(l - 10.95\cdot10^2\cdot\xi); \quad (5)$$

- Taking into account the failure of karst (initial parameters $M_0 = q\cdot l_1^2/2,\ Q_0 = q\cdot l_1$)

$$y(\xi) = y_0\cdot Y_1 + 1095\cdot\varphi_0\cdot Y2 + 1.04\cdot q\cdot l_2\cdot Y_3 + 91.18\cdot10^2\cdot q\cdot l\cdot Y_4 - 998.38\cdot10^4\cdot q\cdot Y_1\cdot\xi; \quad (6)$$
$$\varphi(\xi) = \varphi_0\cdot Y_1 - 0.365\cdot10^{-2}\cdot(0.125\cdot l\cdot Y_2 + 10.95\cdot10^2\cdot Y_3); \quad (7)$$
$$M(\xi) = 0.12\cdot y_0\cdot Y_3 + 130\cdot\varphi_0\cdot Y_4 - 3\cdot10^3\cdot q\cdot(1.5\cdot10^3\cdot Y_1 + 10.95\cdot10^2\cdot Y_2) - 1199.025\cdot10^{\cdot5}q\cdot Y_3\cdot\xi. \quad (8)$$

Here, Y_1, Y_2, Y_3, Y_4 are A.N. Krylov function are widely used in engineering calculations, and therefore their values are made up table for values of ξ argument. A numerical study of deflections, angles of rotation and bending moments, depending on the argument ξ (fig. 4).

Using the obtained expressions, numerical studies were performed to strip foundations of buildings. Calculations showed that the stresses in the cross-sections of the strip foundation under karst holes may some cases increased to 21 times.

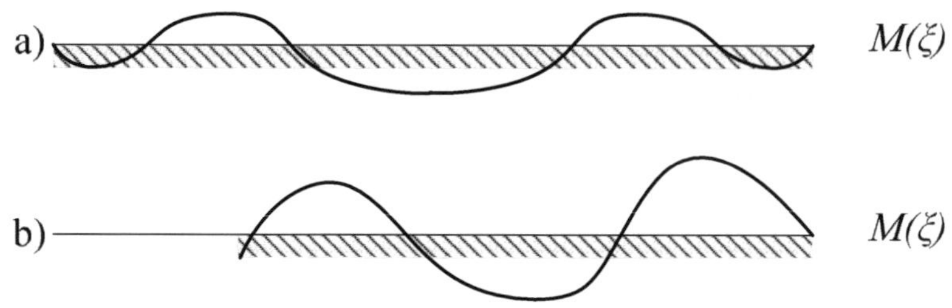

Figure 4. Changings of bending moment and transverse force on length of strip foundation.

3. Conclusion

Analysis of beams on elastic foundation is of great importance in construction practice. The problems associated with the calculation of structures lying on an elastic foundation under karst failures, represent one of the most urgent, complex and interesting problems of structural mechanics.

The paper calculated the stresses and strains in the cross sections of the strip foundation evenly distributed load in the event of karst. The state of stress and deformation of the strip foundation on an elastic foundation depends on the load and the stiffness of the foundation and the elastic foundation. A comparison of extreme stress in the event of karst and without accounting for strip foundation. The computational model of strip foundation - a beam on an elastic foundation. The stresses in the cross-sections of the strip foundation at Karst failure increased from 5 to 21 times. The calculating formulas for determining the parameters of the stress-strain state of strip foundation in any of its cross-section in case of karst and without registration. This calculation method can be applied to the strip foundations of various structures that transmit the load to the ground and to the foundations of dams.

References

[1] Veress M 2010 *Karst Environments* (Berlin: Springer).
[2] Assaad F and Jordan H 1994 *Environmental Geology* **23** 228-237.
[3] 1995 *Guidelines for geotechnical investigations in areas of karst* (Russia, Moscow) 167.
[4] Lodigina N and Sharapov R 2014 Calculation of pile foundations at the karst areas *Proc. of MEACS 2014* (Tomsk, Russia, 16-18 October 2014) 1-4.
[5] Tolmachev V 1986 *Engineering and building development of karst territories* 176.
[6] Dubrovkin V L 1948 Karst study of new processes in seeking to track rail lines *Soviet Geology* **35**.
[7] Sharapov R V, Kuzichkin O R Monitoring of karst-suffusion formation in area of nuclear power plant *Proc. of the IDAACS 2013* **2** 810-813.
[8] Bondarik G K, Pendin V V and Jarg L A 2009 *Engineering Geodynamics* 440.
[9] Parise M and Gunn J 2004 Natural and anthropogenic hazards in karst areas *Natural Hazards and Earth System Sciences*, **23**.
[10] Lodigina N, Sharapov R 2015 Calculation of Grillage Strip Foundation in Area of Karst Collapses Formation *Applied Mechanics and Materials* **770** 723-728.
[11] Lei M, Jiang X and Yu L 1997 Evaluation of karst collapse: theory and methodology *J Geol Hazard Control* **8** 38–42.
[12] Andreo B, Carrasco F, Dura´n J and LaMoreaux J 2010 *Advances in research in karst media* (Berlin: Springer)
[13] Ford D C and Williams P 2007 *Karst Hydrogeology and Geomorphology* (John Wiley and Sons).
[14] Tolmachev V 2007 Issues of environmental impacts of karst in standards on construction in Russia *Environ Geol* **51** 787–790.

[15] Veni G, DuChene H, Crawford N C, Groves C G, Huppert G H, Kastning E H, Olson R and Wheeler B J 2001 Living with Karst: a Fragile Foundation *Environmental Awareness Series*, American Geological Institute.
[16] Rodionov N 1958 *Engineering-geological studies in karst areas.*
[17] Gotman A L and Magzumov R N 2014 Calculation method of pile strip foundations in the formation of karst failure *Vestnik MGTU* **2** 74-83.
[18] Il'ichev V A and Gotman N Z 2006 Evaluation of the base design parameters in the numerical modeling of a structure and a pile base interaction with karst deformations *Proc. of the Thirteenth Danube-European Conference on Geotechnical Engineering* **2** Ljubljana 605-610.
[19] Gotman A L 2008 Analysis of anti-karst foundations for buildings and structures *Soil Mechanics and Foundation Engineering* **45** 23-29.
[20] Gorbunov-Posadnov M and Malikova T 1973 *Construction calculation on elastic base* (Moscow).
[21] Smirnov A 1975 *Matherial resistenses* (Moscow).
[22] Sharapov R V 2014 Indicators for karst processes evaluation *Proc. of the SGEM2014* (17-26 June 2014, Albena, Bulgaria) 519-525.

IOP Conf. Series: Materials Science and Engineering **221** (2017) 012024 doi:10.1088/1757-899X/221/1/012024

Elaborating Surface Topology Control System for Technological Objects

O N Kuzyakov, M A Andreeva and U V Lapteva

Industrial University of Tyumen, 70 Melnikayte Str., 227, Tyumen 625027, Russia

E-mail: onkuzyakov@mail.ru

Abstract. The basics of elaborating topology control system for technological objects are presented in the paper. The main part of the system is measuring block with laser and ultrasonic distance measuring devices, which enables to have technological object scanned with accuracy demanded. The algorithm of scanning technological object, in particular, vertical storage tank, is presented as well. The system elaborated is an integral part of intellectual decision-support system.

1. Introduction

Currently, elaborating surface topology control system for technological objects represents a task. The features required are high sensibility, high level of automatization and universality (ability to scan a range of technological objects). In particular, such system is to be exploited during scanning vertical storage tanks (VST) in order to identify defects (splits, dents, etc.).

According to [17, 18], the periodicity of full VST survey should be 8-10 years. Such survey includes internal side condition control [19], which is held with the use of manual different technologies. However, such technologies involve high labor intensity and high degree of risk for maintenance engineers.

The execution of a range of automated devices is based on a photometrical method with the use of so-called Moire effect [1-6, 20-22], obtained by projecting a grid of alternating light and dark bands on the object investigated. It is followed by the imposition of image of this grid received to another grid located at some distance from the object and rotated by a known angle. The Moire pattern formed this way provides with information about the object topology. The disadvantage of this method is high complexity of processing data and low resulting accuracy.

Thus, the task is to elaborate a surface topology control system for technological objects, which uses up-to-date high-sensible non-contact methods. The device measuring distance or linear displacement is to be the main element of the system. Signal sent by such device allows identifying local defects of the surface of technological object or surface defects in general.

2. System elaboration

An analysis of measuring devices undertaken showed the laser and ultrasound devices are the mostly commonly used. Distance measurement using these devices involves triangulation or time-of-flight method. Laser measuring devices are more sensible than ultrasound ones. Laser devices measurement accuracy is 0.1 mm and the maximum distance to the object is about hundreds of meters, ultrasound has 1 mm and 20 m for the same features.

IOP Conf. Series: Materials Science and Engineering **221** (2017) 012024 doi:10.1088/1757-899X/221/1/012024

A surface topology control system for technological objects (VST in particular) was elaborated according to the analysis mentioned. The measuring devices of the system are ultrasound and laser distance measuring devices. The structure of the system is presented at figure 1.

A measuring block 1 consists of a scanning block 2, which, in turn, has ultrasound and laser distance measuring devices, and microcontroller 3. The measuring block 1 is firmly attached to the mechanism 4 by the metal telescopic bar-bell 5. The mechanism 4 includes two stepper motors, the first of which rotates the bar-bell 5 on its axis and the second displaces it in vertical direction, and stepper motors control block connected to the computer (PC) 6. A measuring block 1 is connected to the PC 6 on the wireless network interface.

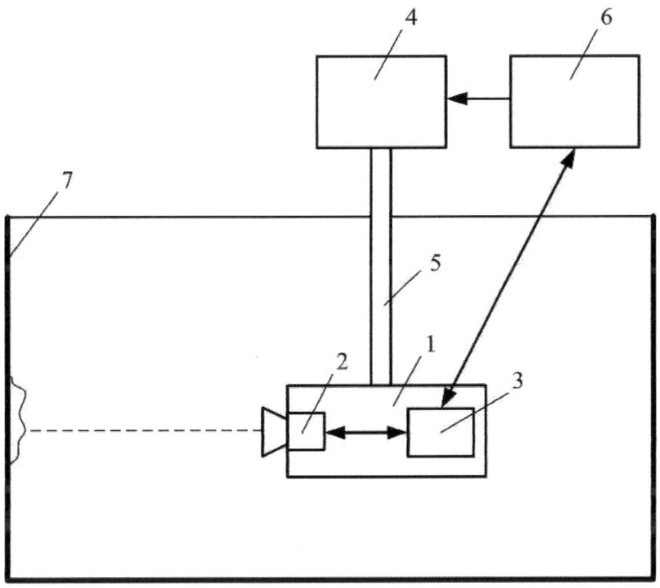

Figure 1. System structure

The system executes as follows: the mechanism 4 displaces the telescopic bar-bell 5 so that the measuring block 1 is level with the highest scanning level. After that the mechanism 4 rotates the bar-bell 5 on its axis in 360 degrees, and the measuring block 1 is switched on from the beginning of the bar-bell 5 rotation. During rotation, the scanning block 2 measures the distance to the VST 7 surface point. The data flows from the scanning block 2 to the microcontroller 3 where data is initially processed according to the program downloaded do the microcontroller 3. The data processed flows from the microcontroller 3 to the PC 6 by means of wireless network interface. The PC 6 executes the major amount of processing and systematizing work according to the program downloaded to it.

After the bar-bell 5 has come full circle, the second stepper motor displaces it a numerous steps down in vertical direction. Then the bar-bell 5 rotates on its axis again and data from the scanning block 2 flows to the microcontroller 3 and finally to the PC 6.

To determine the surface points to which the distance is going to be measured during scanning, the scanning map if formed considering physical parameters of technological object. In case of VST internal side surface the scanning map looks as at figure 2.

The scanning map is formed as follows: horizontal strings, which represent horizontal scanning levels, are put on the side surface model of VST whereby the highest scanning level matches the top border of the model as shown at figure 2. The model height H equals VST height and its length is $2\pi R$ where R is VST radius. The distance between the horizontal scanning levels v equals the value of vertical scanning step.

After scanning levels are formed, scanning points which represent the surface points to which the distance is to be measured are put along the length of each scanning level. The distance between the points h equals the value of horizontal scanning step. The first scanning point of each level matches the left border of VST side surface as shown at figure 2.

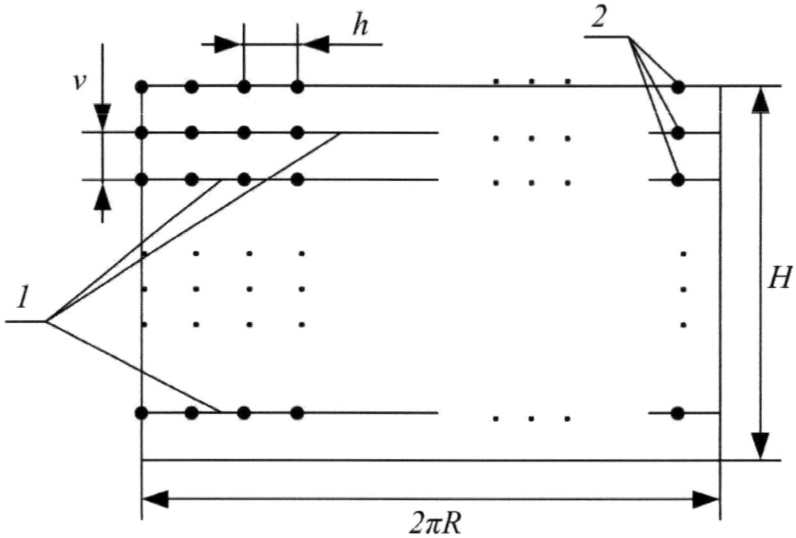

Figure 2. Scanning map formed

Therefore, scanning map of technological object (VST in particular) is a set of scanning points with the distance h (the value of horizontal scanning step) between them, which are put along the length of each scanning level with the distance v (the value of vertical scanning step) between two next levels. The overall number of measurements equals the overall scanning points number K:

$$K = \frac{1000 \cdot H}{v} \cdot \frac{1000 \cdot 2\pi R}{h}, \tag{1}$$

where H is VST height value (m), R is VST radius value (m), v is the value of vertical scanning step (mm), h is the value of horizontal scanning step (mm).

The topology control system for VST internal side surface executes in accordance to algorithm shown at figure 3.

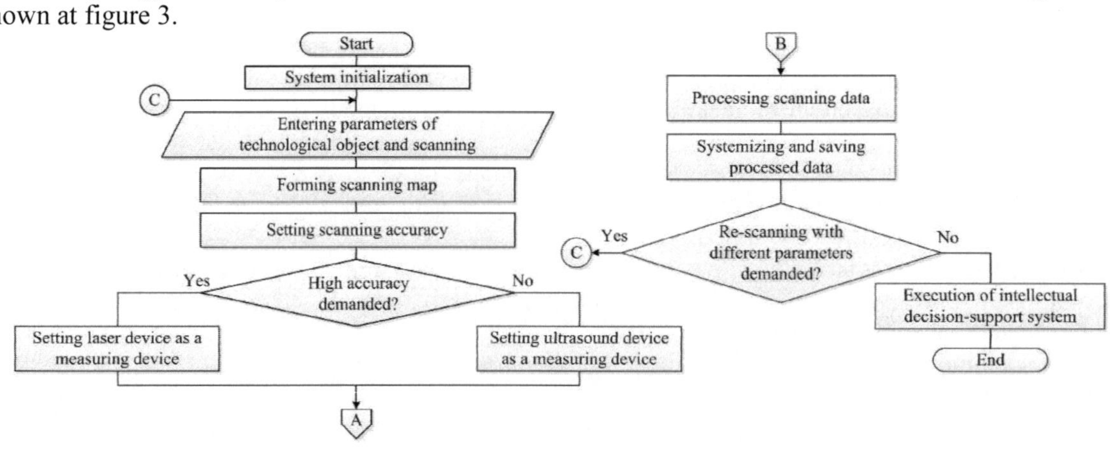

Figure 3. System execution algorithm

IOP Conf. Series: Materials Science and Engineering 221 (2017) 012024 doi:10.1088/1757-899X/221/1/012024

According to the algorithm, system initialization is to be held at the beginning of system execution. After that the following VST parameters and scanning parameters are entered: the height value *H*, the radius value *R*, the horizontal scanning step value *h*, the vertical scanning step value *v*.

The system sets scanning accuracy considering the parameters entered. The decision is made the following way: if at least one of *h* and *v* values is smaller than 0.1 mm, high accuracy is demanded. Otherwise, high accuracy is not demanded.

In accordance with decision-making result on VST scanning accuracy the system sets the measuring device. If high accuracy is demanded, the laser device is to be the means of measuring, otherwise, the ultrasound device is chosen.

After that level-by-level VST scanning starts in conformity with algorithm at figure 4.

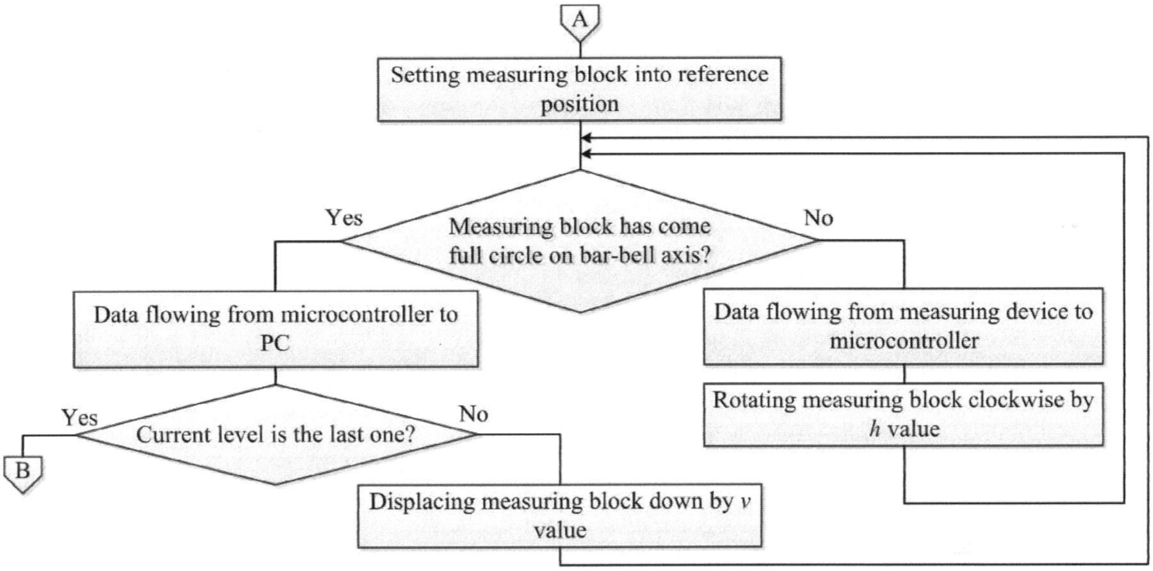

Figure 4. VST scanning algorithm

VST scanning begins with mechanism positioning the measuring block 1, which consists of laser or ultrasound measuring device, into the reference position. This considers the measuring device orientated towards the first scanning point of the highest scanning level.

Each horizontal scanning level is scanned as follows: after the distance to the scanning point is measured, the measuring block rotates clockwise by the horizontal scanning step value *h*. While rotating data flows from the measuring device to the microcontroller, which keeps data until the end of current level scanning. When the mechanism has rotated the measuring block on its axis in 360 degrees (counted from the first scanning point of current level), i.e. scanning of current level finished, the scanning level data flows from the microcontroller to the computer, which keeps it until the end of VST scanning.

Then system checks if current scanning level is the last one. In case there are more levels to be scanned, the measuring block is displaced down by the vertical scanning step *v*, i.e. the measuring module moves down to the next scanning level. Otherwise, VST scanning finishes.

After the end of scanning, processing of VST scanning data executes in accordance to the program kept in PC. Data systematizing and displaying result are executed thereafter.

In case VST scanning with different parameters is demanded, system execution goes back to parameters entering step. Otherwise, intellectual decision-support system [23-24] execution starts. The intellectual decision-support system role is to search a similar precedent in precedent base and to save a new precedent or to adjust it in order to make a final decision [7-16]. A precedent is defined as follows:

$$\text{CASE} = \sum L_{hvi}, M, \tag{2}$$

where $\sum L_{hvi}$ is a set of parameters measured (distances to the scanning points in particular), M is a complex parameter including similarity degree of current set of parameters to precedent from precedent base, precedent extraction algorithm and decision.

After that VST surface topology control system execution suspends.

3. Calculations

In order to evaluate the effectiveness of system elaborated, necessary memory amount and time for technological object scanning were counted (in case of VST).

The memory amount P_h (bits), necessary for storage horizontal scanning level data is given by:

$$P_h = \frac{2\pi R \cdot 1000}{h} \cdot n, \tag{3}$$

where R is VST radius value (m), h is a horizontal scanning step value (mm), n is a microcontroller parity value (n is assumed to be equal 8).

The memory amount P (bits), necessary for storage full VST scanning data is calculated upon the formula:

$$P = K \cdot n, \tag{4}$$

where K is a number of scanning points. Considering equations (1), (3), (4), for VST-100 where H = 6 m, R = 2,365 m, the calculations were held. Results of the calculations are presented in table 1.

Table 1 shows that at some values of set parameters (vertical and horizontal scanning step values) it is more effective to use streaming data while sending scanning data from microcontroller to PC. As an example, for VST-100 considering $v = h = 10$ mm it is effective to send data after the end of each horizontal level scanning as shown in algorithm at figure 4. The system is to make the decision about a proper data flowing strategy considering VST parameters and scanning parameters entered.

Table 1. Memory amount for storage horizontal scanning level data and full vst-100 scanning data

Vertical scanning step, mm	Horizontal scanning step, mm	Memory amount for storage horizontal scanning level data, Kbytes	Memory amount for storage full VST-100 scanning data, Mbytes
5	5	2.90	3.40
10	10	1.45	0.85
15	15	0.39	0.97
20	20	0.29	0.73

The amount of time t, necessary for VST internal side surface scanning is calculated as given by:

$$\mathrm{t} = \mathrm{K} \cdot \mathrm{t_m} + \left(\frac{\mathrm{H} \cdot 1000}{\mathrm{v}} - 1\right) \cdot \mathrm{t_v}, \tag{5}$$

where t_m is data reading periodicity value of the measuring device (msec), t_v is a time of measuring block displacement (t_v is assumed to be equal 1 sec).

IOP Conf. Series: Materials Science and Engineering **221** (2017) 012024 doi:10.1088/1757-899X/221/1/012024

Considering equation (5), VST scanning time is to be calculated for values of vertical and horizontal scanning steps $h = v$ which equal 5, 10, 15 and 20 mm. Results of the calculations are presented in table 2.

Table 2. VST-100 scanning time

Data reading frequency, Hz	Data reading periodicity, msec	Scanning step value ($h = v$), mm	Scanning points number (in each level)	Each level scanning time, sec	Full scanning time, h
200	5.0000	5	2972	14.86	17.83
		10	1486	7.43	4.46
		15	991	4.95	1.98
		20	743	3.71	1.11
250	4.0000	5	2972	11.89	14.27
		10	1486	5.94	3.57
		15	991	3.96	1.58
		20	743	2.97	0.89
300	3.3333	5	2972	9.91	11.89
		10	1486	4.95	2.97
		15	991	3.30	1.32
		20	743	2.48	0.74
400	2.5000	5	2972	7.43	8.92
		10	1486	3.71	2.23
		15	991	2.48	0.99
		20	743	1.86	0.56

The calculations show that increasing data reading frequency and horizontal and vertical scanning step values results in decreasing amount of time necessary for scanning technological object.

4. Conclusion

Thus, a universal automatic system for surface topology control of technological object (VST internal side in particular) elaborated is characterized by lack of human factor influence. Laser and ultrasound measuring devices, which are the main parts of the system, provide with an opportunity to undertake the surveys with accuracy demanded. Decision support of choosing necessary scanning accuracy and identifying the nature of the defects is realized in the system as well.

References

[1] Chiany F P 1979 Moire Methods of Strain Analysis *Exp. Mech.* **19** pp 230-308

[2] Post D 1967 Analysis of Moire Fringe Multiplication Phenomena *Appl. Opt.* **11** pp 1938-1942

[3] Guild J 1956 *The Interference systems of Crossed Diffraction Gratings* (Oxford: Clarendon Press) p 152

[4] Durelli A J, Sciammarella C A, Parks V J 1973 *Interpretation of Moire Patterns* J. Eng. Mech. Div., Proc. ASCE **71**

[5] Durelli A J, Rajaiah K 1980 Determination of Strains in Photo-Elastic Coatings *Exp. Mech.* **20** pp 57-64

[6] Redner A S 1980 Photoelastic Coatings *Exp. Mech.* **20** pp 403-405

[7] Schank R C, Abelson R P 1977 *Scripts, plans, goals and understanding* (Erlbaum, Hillsdale, New Jersey, US) p 248

[8] Simpson R L 1985 *A computer model of case-based reasoning in problem solving: an*

IOP Conf. Series: Materials Science and Engineering **221** (2017) 012024 doi:10.1088/1757-899X/221/1/012024

investigation in the domain of dispute mediation Technical Report GIT-ICS-85/18
[9] K.J. Hammond 1986 *CHEF: A model of case-based planning* In Proc. American Association for Artificial Intelligence, AAAI-86 pp 267-271
[10] Sycara E P 1987 *Resolving adversial conflicts: An approach to integrating case-based and analytic methods* Ph.D Thesis Technical Report No. GIT-ICS-87/26
[11] Koton P 1989 *Using experience in learning and problem solving* Ph.D. Thesis MIT/LCS/TR-441
[12] Althof K-D, Auriol E, Barlette R, Manago M 1995 A review of industrial case-based reasoning tools *AI Intelligence*
[13] Information on http://www.iccbr.org/
[14] Aamodt A, Plaza E 1994 Case-based reasoning: foundational issues, methodological variations, and system approaches *AI Communications, IOS Press* **7** pp 39-59
[15] Alterman R 1989 *Panel discussion on case representation* In Proceedings of the Second Workshop on Case-Based Reasoning
[16] David B S 1991 *Principles for case representation in a casebased aiding system for lesson planning* In Proceedings of the Workshop on Case-Based Reasoning
[17] Guidance document 19.100.00-KTH-545-06 *Ultrasoung control of side surface and welded joins while exploitation and renovation of vertical storage tanks* (in Russian)
[18] Guidance document 153-39.4-078-01 *Regulars of operation of oil-trunk pipelines and oil bases tanks* (in Russian)
[19] State standard 8.570-2000 ICG *Vertical cylindrical storage tanks. Audit methodology* (in Russian)
[20] Kuzyakov O N, Kucheryuk V I 2002 *Methods and means of measuring surface topology, displacements and deformations* (Tyumen: TSOGU) p 172 (in Russian)
[21] Kucheryuk V I, Kuzyakov O N, Dubatovka U V 2005 Surface topology identifying device by using Moire method *Patent RF* 2267087
[22] Kucheryuk V I, Mishenev A A 2012 Tanks defects control device with Moire method *Patent RF* 2454627
[23] Bashlykov A A 2016 The basics of projecting informational systems for intellectual decision-support *Automatization, telemechanization and communication in oil industry, Moscow: OJSC "VNIIST-Neftegazproekt"* **11** pp 12-24 (in Russian)
[24] Bashlykov A A 2014 Analysis of organization intellectual information decision-support in emergency situations *Automatization, telemechanization and communication in oil industry, Moscow: "VNIIOENG"* **8** pp 3-9 (in Russian)

IOP Conf. Series: Materials Science and Engineering **221** (2017) 012025 doi:10.1088/1757-899X/221/1/012025

Problem-Oriented Corporate Knowledge Base Models on the Case-Based Reasoning Approach Basis

I N Gluhih[1, a], R K Akhmadulin[2]

[1] Tyumen State University, 6Volodarskogost., Tyumen, 625003, Russia
[2]Tyumen Industrial University, 38 Volodarskogost., Tyumen, 625000, Russia

E-mail: [a] igluhih@utmn.ru

Abstract. One of the urgent directions of efficiency enhancement of production processes and enterprises activities management is creation and use of corporate knowledge bases. The article suggests a concept of problem-oriented corporate knowledge bases (PO CKB), in which knowledge is arranged around possible problem situations and represents a tool for making and implementing decisions in such situations. For knowledge representation in PO CKB a case-based reasoning approach is encouraged to use. Under this approach, the content of a case as a knowledge base component has been defined; based on the situation tree a PO CKB knowledge model has been developed, in which the knowledge about typical situations as well as specific examples of situations and solutions have been represented. A generalized problem-oriented corporate knowledge base structural chart and possible modes of its operation have been suggested. The obtained models allow creating and using corporate knowledge bases for support of decision making and implementing, training, staff skill upgrading and analysis of the decisions taken. The universal interpretation of terms "situation" and "solution" adopted in the work allows using the suggested models to develop problem-oriented corporate knowledge bases in different subject domains. It has been suggested to use the developed models for making corporate knowledge bases of the enterprises that operate engineer systems and networks at large production facilities.

1. Introduction

In recent decades one of the important directions of production processes automation development and activity management of enterprises of different areas has become a development of corporate knowledge systems [1-7]. Corporate knowledge bases (CKB) allow saving and use of the knowledge, available to the enterprise collective, for many tasks: from new product design and production to provision of commercial activities and managerial decision making.

Despite the urgent character of knowledge management systems there are still no single technological solutions to build CKB. An approach, popular today, uses content management systems(CMS), where knowledge can be represented in the form of texts, images and other content components created by corporation employees [7-9]. A corporate knowledge base obtained as a result of such an interaction, is represented by a network with a variety of semantic ties and associations which do not often lend themselves to precise definitions.

Due to the diversity of ties and content components and lack of the clearly defined task-oriented structure it is difficult to search and use knowledge to solve specific problems that may arise before users at an enterprise. As a consequence there is a reduction of CKB effectiveness and doubt about the usefulness of the knowledge management system as a whole.

This article suggests a new conceptual model – a model of problem-oriented corporate knowledge bases (PO CKB) – which allows making the system of enterprise knowledge more constructive and task oriented.

In PO CKB corporate knowledge (content) are arranged around problem situations that arise during the corporation (enterprise) activities.

These situations are characterized by the fact that a) in the current situation, the obvious and only mode of action is missing; b) the situation requires creative decision-making from the staff, which is associated with options generation and analysis; c) the execution of these decisions involves a set of actions and documents and is carried out in accordance with definite, possibly various scenarios.

The basic idea of CKB, oriented to problem situations, is to create a system that will provide the knowledge for making and implementing decisions in such situations (ready-made decisions, actions scenarios, regulatory documents, specifications, cost estimates, contract templates, etc.).

Problem-oriented CKB allows not only to support the decision makers, but to use the corporate knowledge and experience for training, skill upgrading and commissioning of new employees, as well as to make corporation activity results retrospective analysis.

Further on a domain-independent knowledge representation model to implement this concept has been suggested and a generalized PO CKB structural chart has been developed.

2. Method and models

2.1. Case-based reasoning

A theoretical basis of building problem-oriented CKB is the case-based reasoning method in intelligent systems [10, 11].

The basic idea of CBR is to solve emerging problems using the decisions already known from the past experience which were previously used in practice or were suggested by experts for similar problems.

The knowledge bases components in CBR systems are cases that are stored in a case-base (CB) (library).

Each case is represented by the couple <Situation, Solution>. Under the situation a formalized description of a problem to be solved is meant.

In a general case anything that can be used or recommended in this situation is represented as the solution.

The work of the CBR-system includes the following major stages [12]:

- «retrieve» – search and selection of the case which is most suitable for solving a current problem in CB. The search key is the description of the current user situation Sit, which is compared with situations S from the case-base. At the end of this stage one or more S* which are most similar to situation Sit are selected;

- «reuse» – retrieving and reusing of solution R*, which enters the same case as S*.

If necessary, solution R* is adapted to a new situation. If a new solution for a new problem was obtained, after the results auditing the new case is added to the CB.

Case base reasoning to find solutions has shown its effectiveness in different problem areas. For example, this method is used for traffic control at crossroads [13], in software development support systems [14], to identify damages and predict mechanical systems states [15, 16], for text classification [17], in medical systems [18-21] to predict the cost of construction [22, 23], etc.

The assessment of closeness of situations Sit to S at the stage of «retrieve» is one of the most important tasks in CBR. The solution to this problem depends on the chosen method of formalized representation of situations. There are two main approaches to compare situations [11] – selection of the most similar situations based on distance measurement in the parameter space, and selection of situations based on the decision tree.

The work [24] suggests a decision tree modification – the situations tree (ST), which further will be complemented with new elements specifically for knowledge representation in PO CKB.

The situations tree is a binary ratio graph $\rho(S^j,S^{j+1})$ "type-subtype", where situation S^{j+1} at the (j+1)-th level is a description elaboration of situation S^j located at the previous, j-th level. This means that the set of attributes – indicators of situation A^j are included in set A^{j+1} of description of situations at the next level.

In this case two situations, S^j_i and S^j_r, of the same level differ in the regions of the allowed values in the parameters space $A^j_i \in D^j_i$, $A^j_i \in D^j_r$, where $D^j_i \neq D^j_r$.

The peculiarity of the situations tree is that its tops at each level are not abstract sets of variables, but correspond to some practically relevant types of situations. Each of them can match with its decision. The decisions at the upper levels have a higher level of generalization and are less detailed than the decisions at lower levels, but true for all situations, subtypes of the given situation of the top level.

2.2. PO CKB model of knowledge

In CBR-based problem-oriented CKB a case-base is a knowledge base. A formalized knowledge model is designed for domain-independent case view, the interrelations between them, as well as for case search and selection. When building such a model in PO CKB, we consider the following features:

- in enterprise activity there can be identified typical, recurring problem situations. They are characterized by a common set of features, some of which are relevant to the classification of situations, and others determine the problem context (specifies the place and time of occurrence, external conditions, decision executer, etc.);
- in these situations the solutions and the results can differ. The solutions effectiveness is estimated by experts based on subsequent analysis of the results;
- in the corporate knowledge base examples of situations and their solutions should be saved as well as some generalizations, i.e. descriptions of typical situations and typical solutions, which include activity templates, documents, typical scenarios, etc.

Thus, situations-examples will be distinguished from typical situations in CKB. A typical situation is characterized by field D of possible values of the attributes vector, and situation-example – by a point in field D.

The knowledge model in PO CKB, implemented based on the situations tree, should represent both typical situations and situations-examples related to them.

In the particular case the situations-examples are matched with the lower levels and the upper levels are used to obtain generalized solutions and a sequential search according to the values of attributes (see Figure 1). In a general case, many examples can be matched with each top S.

Taking into consideration the above mentioned, case C in the case base is represented by the following set of components associated with the tops of the situations tree:

$C = \{S, R <SE, RE, P>k \mid k = 1,..., M\}$,

where S is a description of a typical situation, given through a set of allowed values of attributes D = {d1, d2, ..., dn}, where n is the number of attributes-indicators of this situation;

R is a typical solution for this situation, which includes the following:

R1 is an imperative part – a guide to an action in this situation, and answers the question "What to do?"

R2 is a regulatory and guidance part, and answers the question "How to do?"

R3 is a reference part, additional information which may be required for decision implementation (reference data, illustrations, models, etc.), "Through what to do?";

R4 is a recommendatory part that contains experts' recommendations of how to implement the proposed decisions;

<SE, RE, P >is an example of a situation of this type, the decision made in this situation and the result obtained, respectively.

M is a number of examples in a case.

IOP Conf. Series: Materials Science and Engineering **221** (2017) 012025 doi:10.1088/1757-899X/221/1/012025

To describe situations-examples in PO CKB and for advanced search capabilities in CB elaboration of the set of attributes is desirable [4]:

SE⇒ {A, AC, T, I }, where A is a vector of situation attributes-indicators, which allows relating the situation to the given type $A = (a_1, a_2, \ldots, a_n) \in D$; AC are additional attributes to describe the context which are not used in the identification of the type; T is a text meta-description of a situation; I – a graphic image of the situation (photo, illustration, picture cards, etc.).

In PO CKB situation Sit stands for a user situation – a certain situation interesting for the user. It could be a current problem situation, which requires an immediate solution, or a hypothetical situation, interesting for the purpose of teaching, or a situation from a corporation's past experience.

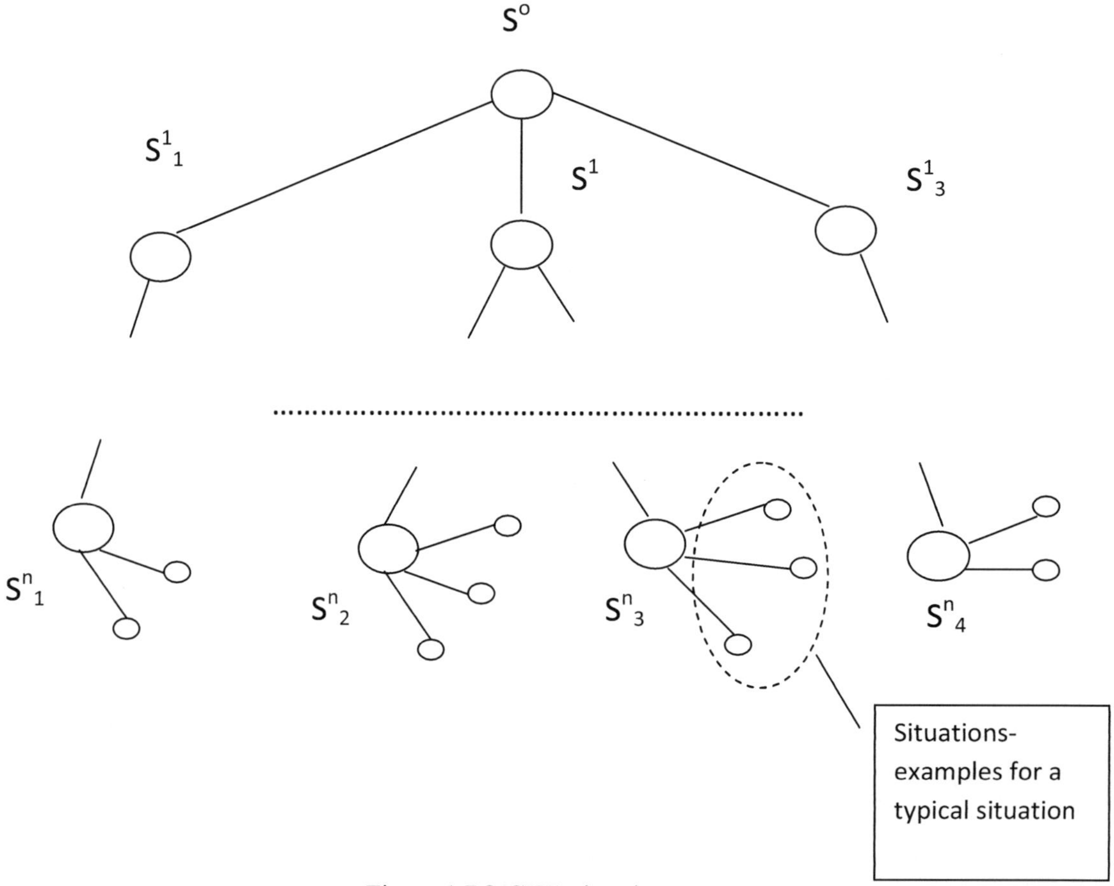

Figure 1.PO CKB situations tree

In a general case the user lacks not only an accurate description of Sit but also full information about those attributes which values need to find out to compare with the examples in CB. In case-based reasoning this gives rise to the problem of Sit identification, which is complicated by an increase in the volume of CB and by the occurrence of situations with different sets of attributes-indicators. It can become a serious obstacle to the practical implementation of CBR systems.

The suggested model allows overcoming this difficulty by a sequential search through the situations tree from the top down. Here each top S of the tree represents the set of attributes that should be inquired for further movement. Thus, alongside the search on the tree the identification of Sit is performed, and in this case only those attributes are inquired that have a value for this type. Eventually, the search leads to the desired top S* with a set of allowed values D*, for which $A(Sit) \in D^*$ is performed, where A(Sit) is a vector of the user situation attributes values.

IOP Conf. Series: Materials Science and Engineering **221** (2017) 012025 doi:10.1088/1757-899X/221/1/012025

2.3. Structural model

Based on the developed knowledge model for case views, there has been suggested a generalized structural chart of PO CKB (Figure 2). Since at existing plants, as a rule, there are various databases and documents, PO CKB is implemented as an add-in to the existing corporate accounting systems and content management systems. The descriptions of the situations-examples and their solutions include references to the known data and the documents or procedures necessary to be obtain in the existing systems.

As shown in figure 2, apart from the end-user a cases base editor is required to replenish CB. The system provides the following basic modes:

- search on query: search of a typical situation and (or) subset of situations-examples by attribute values (search on situations tree); full-text search by textual description T, search by the graphic image of situation I;
- cases base training (replenishment). With emergence of new situations and new solutions a new case is added to the cases base or the existing case is updated. If this new situation-example is an example of a situation type already present in CB, then the existing case is added. If this is a new type situation, the description of a typical situation is added and a case with new examples is formed. In this case a new top should be added to the situations tree, and its location will be determined by the ties of the new typical situation with the already existing ones.

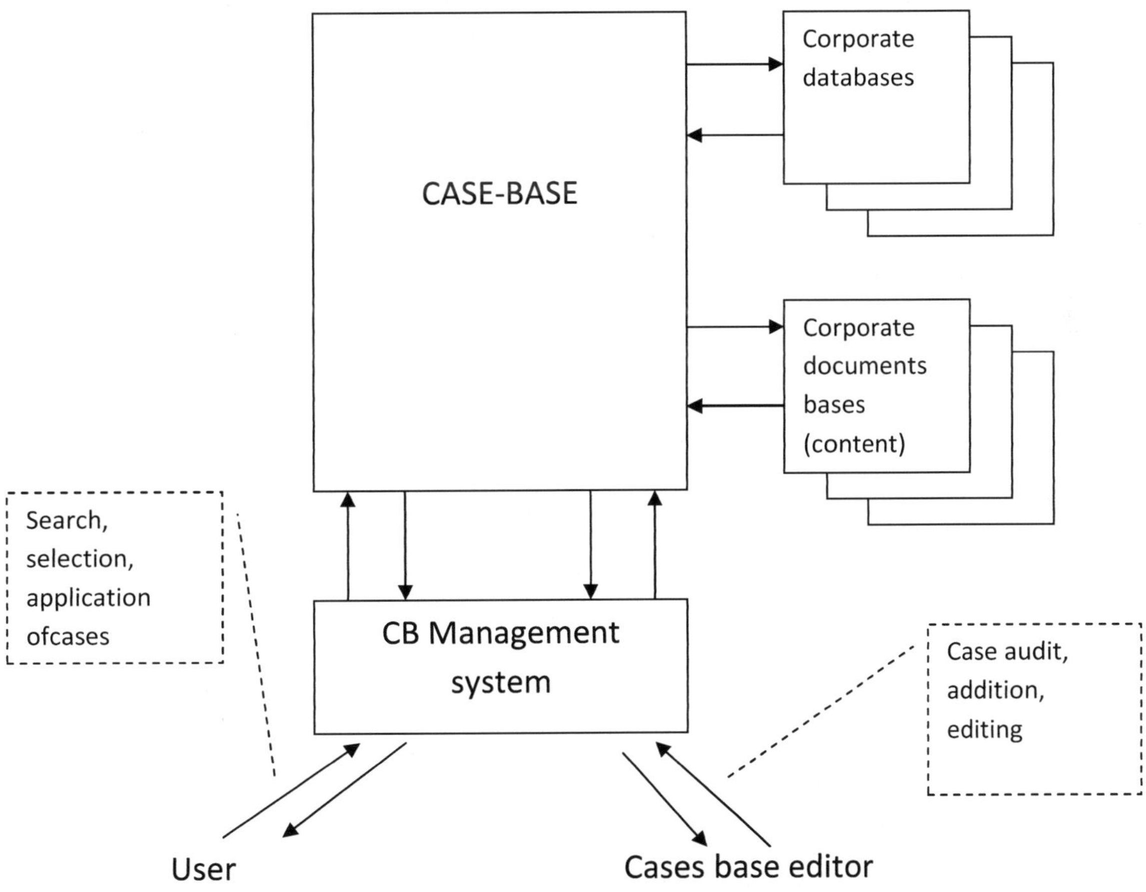

Figure 2.PO CKB structural model

A monitoring subsystem and current situation analysis can be added to the structural model, which can automatically identify the situations that require interference by a given set of parameters coming from the accounting systems. This provides another mode of operation – "consulting system mode» in which the end-user will get ready-made cases to solve an arising problem.

2.4. Problem-oriented knowledge bases to operate engineer systems and networks.
Creation of enterprises corporate knowledge bases to maintain engineer systems of buildings and structures (heat, water and electricity supply, etc.) is a promising application of the developed models in production activities. Operation of engineer systems at large and especially distributed plants is related to complexity and diversity of the equipment, a high probability of emergency situations, and great influence of the service quality on the plant life. PO CKB of an operating organization contains segments – situations trees for each type of the engineer systems. On the upper levels a situations tree divides a number of situations into two classes: S^1_1 - regular situations that require routine maintenance work; S^1_2 - emergency situations which require urgent organizational and technical measures to eliminate accidents and their consequences.

On the lower levels of a situations tree every top S corresponds to cases C, which contain a description of the updated typical situations, the solution components R required to perform routine maintenance or emergency works, as well as examples of situations SE, activities RE and the results of their performance P that took place in practice.

Occurrence of CKB allows consolidating and using the enterprise experience for both effective elimination of emergency situations and performing routine maintenance, as well as making an analysis of the staff activities.

3. Conclusion
In this work, we have suggested a concept of problem-oriented corporate knowledge bases, where knowledge is arranged around problem situations arising in a corporation, and is used to support decisions making and execution in these situations.

To develop a model of knowledge representation in PO CKB, we have used a case-based reasoning approach. As a result of the study a case view structure has been suggested. It contains descriptions of a typical situation, standard solutions, as well as specific situations-examples and solutions-examples. For storage arrangement and case search a knowledge representation model based on the situations tree and invariant to subject domain, has been suggested.

The universal interpretation of terms "situation" and "solution" admitted in the work allows use of the suggested models for the development of PO CKB and support of decisions in various subject domains. One of the promising directions is creation of corporate knowledge bases to maintain and operate engineer systems in buildings and structures of large productive facilities.

Apart from the very decision support database cases can be used for retrospective analysis of the corporation activities, staff training and expansion of corporation best practices.

References
[1] Debora Suryawan, E. Putra and A. Pratiwi 2015 Knowledge management strategy model for small to medium enterprises *International Conference on Science in Information Technology (ICSITech)*, Yogyakarta, 2015, pp. 219-224. DOI: 10.1109/ICSITech.2015.7407807

[2] Roberto Cerchione and Emilio Esposito 2017 Using knowledge management systems: A taxonomy of SME strategies *International Journal of Information Management* Volume 37, Issue 1, Part B, pp. 1551–1562. DOI: 10.1016/j.ijinfomgt.2016.10.007

[3] Gluhih, I.N., Piankov, V.N. andZabolotnov, A.R. 2002 Situational models in corporative knowledge base on geological-technical measures' know-how*Neftyanoehozyaystvo– Oil Industry* 6p. 45

[4] H. Ping Tserng, Meng-Hsueh Lee, Shang-Hsien Hsieh, and Hsiang-Ling Liu 2016 The measurement factor of employee participation for Knowledge Management System in engineering consulting firms *Journal Of Civil Engineering And Management* Vol. 22 ,Iss. 2, pp. 154-167. DOI: 10.3846/13923730.2014.897963

[5] Edgar Serna M., Oscar Bachiller S., Alexei Serna A. 2017 Knowledge meaning and management in requirements engineering *International Journal of Information Management* Volume 37, Issue 3, pp. 155–161. DOI: 10.1016/j.ijinfomgt.2017.01.005

[6] Rainer Weinreich, Iris Groher 2016 Software architecture knowledge management approaches and their support for knowledge management activities: A systematic literature review *Information and Software Technology* Volume 80, pp. 265–286. DOI: 10.1016/j.infsof.2016.09.007

[7] Shan Wana, DongboLia and James Gaob 2016 Exploring the Advantages of Content Management Systems for Managing Engineering Knowledge in Product-service Systems *Procedia CIRP*Volume 56, pp. 446–450. DOI: 10.1016/j.procir.2016.10.087

[8] Jaffar A. Alalwana, Manoj A. Thomasb, H. Roland Weistroffer 2004 Decision support capabilities of enterprise content management systems: An empirical investigation *Decision Support Systems* Volume 68, pp. 39–48. DOI: 10.1016/j.dss.2014.09.002

[9] Benjamin Wehner, Christian Ritter, Susanne Leist 2017 Enterprise social networks: A literature review and research agenda *Computer Networks* Volume 114, pp. 125–142. DOI: 10.1016/j.comnet.2016.09.001

[10] J.L. Kolodner 1992 An Introduction to Case-Based Reasoning*Artificial Intelligence Review* vol. 6, 1992, pp. 3-34.

[11] I.D. Watson and F. Marir 1994Case-based reasoning: A review*The Knowledge Engineering Review* vol. 9, num. 4, pp. 355-381.

[12] Ramon López De Mántaras, David Mcsherry, Derek Bridge, David Leake, Barry Smyth, Susan Craw, BoiFaltings, Mary Lou Maher, Michael T. Cox, Kenneth Forbus, Mark Keane, AgnarAamodtand Ian Watson2005 Retrieval, reuse, revision, and retention in case-based reasoning*The Knowledge Engineering Review* Vol. 20:3, 215–240. DOI: 10.1017/S0269888906000646

[13] Ali Louati, SabeurElkosantini, Saber Darmoul, LamjedBen Said2016 A Case-Based Reasoning System to Control Traffic at Signalized Intersections. *IFAC-PapersOnLine* 49-5, pp. 149–154.

[14] Rodrigo G. C. Rocha, Ryan R. Azevedo, Ygor Cesar Sousa, Eduardo de A. Tavares, Silvio Meira 2014 A case-based reasoning system to support the global software development*18th International Conference on Knowledge-Based and Intelligent Information & Engineering Systems - KES2014. Procedia Computer Science*2014. №35. Pp. 194 – 202

[15] T Olsson, P Funk 2012Case-based reasoning combined with statistics for diagnostics and prognosis*25th International Congress on Condition Monitoring and Diagnostic Engineering IOP Publishing. Journal of Physics: Conference Series* 364 012061 DOI:10.1088/1742-6596/364/1/012061

[16] L.E.Mujica, J. Vehı, J. Rodellar, P. Kolakowski 2005A hybrid approach of knowledge-based reasoning for structural assessment*Institute of physics publishing smart materials and structures*, 14, pp.1554–1562, DOI:10.1088/0964-1726/14/6/048

[17] Y. Li, S.C.K. Shiu, S.K. Pal, J.N.K. Liu. 2006A rough set-based case-based reasoner for text categorization International *Journal of Approximate Reasoning*, 41, pp. 229–255

[18] Gómez-Vallejo H.J., Uriel-Latorre B., Sande-Meijide M., Villamarín-Bello B., Pavón R., Fdez-Riverola F., Glez-Peña D. 2016Case-based reasoning system for aiding detection and classification of nosocomial infections *Decision Support Systems*Vol. 84 Pp. 104-116.

[19] Douali, N.a , De Roo, J.b, Jaulent, M.-C.a2012Clinical Diagnosis Support System based on Case Based Fuzzy Cognitive Maps and Semantic Web *24th Medical Informatics in Europe Conference, MIE 2012; Pisa; Italy; 26 August 2012 through 29 August 2012* Volume 180, 2012, Pp. 295-299.

IOP Conf. Series: Materials Science and Engineering **221** (2017) 012025 doi:10.1088/1757-899X/221/1/012025

[20] López, B., Pous, C., Gay, P., Pla, A., Sanz, J., Brunet, J. 2011 EXiT CBR: A framework for case-based medical diagnosis development and experimentation *Artificial Intelligence in Medicine* Volume 51, Issue 2, Pp. 81-91.

[21] Sreeparna Banerjee, Amrita Roy Chowdhury 2015 Case Based Reasoning in the Detection of Retinal Abnormalities using Decision Trees*Procedia Computer Science* 46, pp. 402 – 408.

[22] Mohamed M. Marzouk, Rasha M. Ahmed. 2011 A case-based reasoning approach for estimating the costs of pump station projects *Journal of Advanced Research* 2, pp. 289–295.

[23] Naderpajouh,N.and Afshar,A. 2008 A case-based reasoning approach to application of value engineering methodology in the construction industry *Construction Management andEconomics*№ 26(4). Pp.363–372.

[24] Ruslan K. Akhmadulin, Igor N.Gluhih, Ivan Y. Karyakin 2016An object-oriented model of case-based reasoning system using situations tree*10th International Conference on Application of Information and Communication Technologies (AICT). 12-14 October 2016, Baku, Azerbaijan. Conference proceedings* P.124-127